Natural History Museum Senckenberg

Guide

3., revised edition
with 56 colour photos, 275 drawings and 17 maps

Kleine Senckenberg-Reihe Nr. 2
Edited by Prof. Dr. Willi Ziegler
Senckenberg Nature Research Society
Senckenberganlage 25
D-6000 Frankfurt am Main 1, West-Germany

Verlag Waldemar Kramer, Frankfurt am Main

All colour photographs are from Erwin Haupt, except for the following: pages 14, 22, 56, 60, 78, 111, 149, 184 Jürgen Sparr; pages 101, 102 Jens L. Franzen; pages 215, 217 Manfred Grasshoff; page 218 Wolfgang Klausewitz.

CIP-Titelaufnahme der Deutschen Bibliothek

Natural History Museum Senckenberg : guide / [Ed.-in-chief : Willi Ziegler. Engl. transl. of the 28th German ed. by Neil Beckhaus. . .] – 3., rev. ed. – Frankfurt am Main : Kramer, 1988
(Kleine Senckenberg-Reihe der Senckenbergischen Naturforschenden Gesellschaft ; Nr. 2)
Dt. Ausg. u.d.T. : Naturmuseum Senckenberg
2. Aufl. u.d.T. : Senckenbergische Naturforschende Gesellschaft : Senckenberg Natural History Museum of the Senckenberg Natural History Research Society
ISBN 3-7829-1109-1
NE: Ziegler, Willi [Hrsg.]; Senckenbergische Naturforschende Gesellschaft: Kleine Senckenberg-Reihe der . . .

Editor-in-chief: Prof. Dr. Willi Ziegler
Responsible Editor: Dr. Gerhard Winter
ISBN 3-7829-1109-1
English translation of the 28th German edition by Neil Beckhaus, Crosbie Matthews (†), Willi Ziegler
Auslieferung für den Buchhandel:
Verlag Waldemar Kramer, 6000 Frankfurt am Main 60
Schriftleitung: Dr. Gerhard Winter
Grafische Gestaltung: Renate Klein-Rödder
Zeichnungen und Pläne: Eveline Junqueira, Renate Klein-Rödder, Hermann Schäfer, Wilhelm Schäfer, Jürgen Wirth
Farblithos: Keim Klischees, Langen
Schwarz-weiß Lithos: Volkmar Thier
Satz: sg Satzservice GmbH, Frankfurt
Druck: Druckerei Kempkes, Gladenbach

Contents

The "Senckenberg" Past and Present 5

Ground Floor (Map and Key to Galleries) . . . 10/11
Outdoor Exhibits (Room 1) 12
Entrance Hall (Room 2) 13
Room 3, Fossils in Mythology (Planned), Temporary Exhibitions, Dinosaur Track . . . 18
Room 4, Ancient Egyptian Animal Symbols . . 20
Room 5, Dinosaur Hall (Central Hall) 24
Room 6, Fossil Vertebrates Adapted to Aquatic Life 32
Room 7, Rocks and Minerals, Fossil Invertebrate Animals 44
Room 8, Fossil Mammals, Archaeopteryx . . . 63
Room 9, Trachodon Cellar 71
Room 10, Whales and Elephants 80
Room 12–13, Origin of Man 91
Room 14–15 (Exhibitions planned) 110

First Floor (Map and Key to Galleries) 112
Room 101–102 Mammals 113
Room 103, Mammals (Provisional) 122
Room 104, Reptiles and Amphibians (Provisional) 124
Room 105–107, Birds 138
Room 108, Paleobotany and Botany (Planned) 165

Second Floor (Map and Key to Galleries) . . . 171
Room 202, Mollusks 172
Room 203, Insects 176
Room 204, Crustaceans and Chelicerata 199
Room 205, Dioramas (Habitat Groups) 213
Room 205, Marine Invertebrates 215
Room 207–208, Fish 220

History of the Senckenberg Nature Research Society 242

Notes

☐ Number of the rooms in different colours (Ground Floor: Green; First Floor: Blue; Second Floor: Red)

⌋ Number of cases (beside the text column)

2nd FLOOR

1st FLOOR

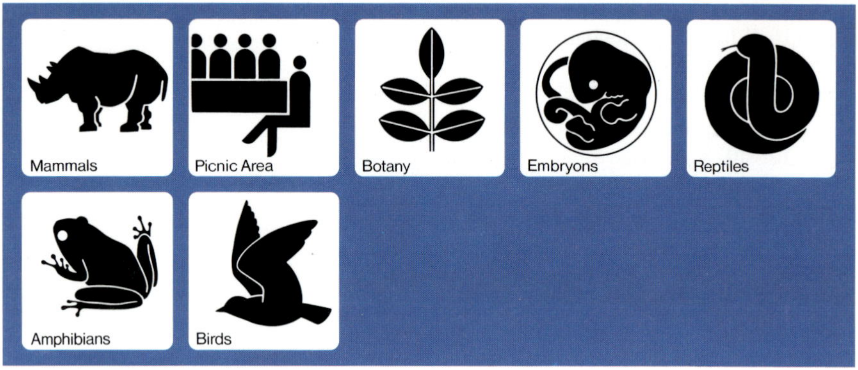

GROUND FLOOR

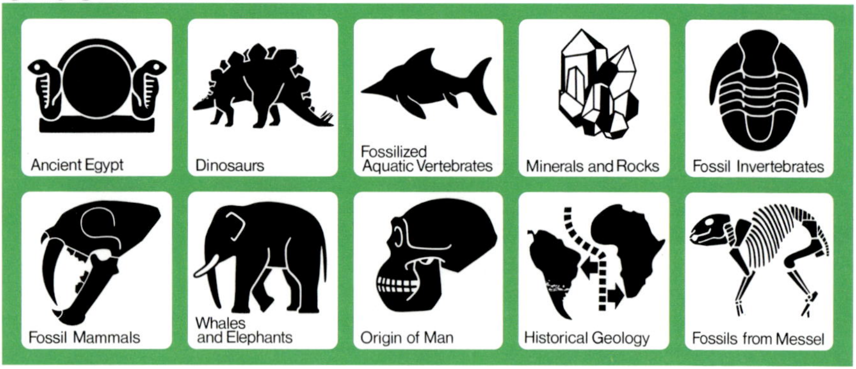

The “Senckenberg” Past and Present

The Senckenberg is the most comprehensive Natural History Research Museum of the Federal Republic of Germany, because of its exhibits, scientific institutions and numerous publications. Besides the grandiose museum with many spectacular and valuable displays, it also houses a research institute with important scientific collections in the fields of zoology, anthropology and paleoanthropology, geology and paleontology, botany and paleobotany, marine geology and marine biology as well as ecology, limnology and landscape research. In addition, the Senckenberg Nature Research Society publishes 11 series of scientific and popular publications which range from a monthly magazine via scientific journals and monographies up to special professional books. Through the unique collections, the research in that broad spectrum of scientific fields and its large series of publications, the Senckenberg works on a grand global scale and compares with the leading research museums of the world. Already in its early history the Senckenberg became world famous. After the famous German poet J. W. Goethe had appealed to the citizens of his home city Frankfurt in 1815 to found for themselves a nature research society and establish a nature museum, the Senckenberg Nature Research Society was brought to life by a group of Frankfurt citizens under the leadership of the physician Dr. P. H. J. Cretzschmar in 1817. Only four years later in 1821 the “Museum Senckenbergianum” was opened, then situated at the Eschenheimer Turm. Immediately thereafter, a famous Senckenberg researcher, Eduard Rüppell, started expeditions leading him to Northeast Africa, Arabia and to the Red Sea (1822 to 1827 and 1830 to 1834). Even in its early days, the Senckenberg Museum received from these expeditions scientific collections so tremendously rich that at the end of the first decade after its foundation it ranked among the great museums of the world. Ever since passionate nature research has been carried out in the Senckenberg Museum.

In the second half of the last century the Senckenberg Museum functioned as a scientific academy as there was not yet a university in the city of Frankfurt. The geological, mineralogical, botanical and zoological university institutes were founded by the Senckenberg Nature Research Society and transferred to the university in 1914, the year of university foundation, and for many years Senckenberg scientists were full professors as heads of the university departments.

The new Senckenberg Museum, opened in 1907, was built in the center of the area that later became the university campus. This spatial and personal proximity of the Senckenberg Museum to the new university led to a close relationship between Senckenberg and the university as well as to academic training for the decades to come.

At about that time, the paleontologist Professor Dr. F. E. Drevermann endeavoured to reform the Senckenberg Museum as a place of broad popular education, and the zoologist, Professor O. L. zur Strassen, director of the Senckenberg Museum and of the Zoological Institute of the University until 1933, gave a strict scientific framework to the museum exhibition, particularly in the fields of biological behaviour and ecology.

Although the time between the two World Wars was very difficult economically for the Senckenberg Nature Research Society, two other of its famous scientists set several very important scientific impulses within that very time: The geologist Professor Dr. R. Richter, director of the Senckenberg Museum and at the same time of the Geological Institute of the University between 1933 and 1945, had already founded the Research Institute for Marine Geology and Marine Biology called "Senckenberg at Sea" in Wilhelmshaven in 1928 and introduced two completely new research fields, namely actuogeology and actuopaleontology on the basis of the philosophy of uniformitarianism, i. e. the present is the key to the past. Both research fields still make up the central research areas of this Senckenberg Institute on the North Sea today. R. Richter, himself a very famous paleontologist and sedimentologist, made the research institute of the Senckenberg Museum a geological-paleontological center of international significance.

Professor Dr. R. Mertens, zoologist and director of the Senckenberg Museum from 1945 to 1959, not only directed the restoration of the various buildings that had been heavily damaged during the Second World War, he also set new impetus to the zoological research in that he expanded research on ecology and behavioural biology. In 1954 the Senckenberg institutions, from now on called Forschungsinstitut (Research Institute) and Naturmuseum (Nature Museum), were included in the so-called "Königstein Treaty", established by the States of the Federal Republic for the benefit of scientific institutions, the work of which was regarded to be superregional.

Until then, Senckenberg and its scientific and museal institutions had been financed by a small annual financial donation of the city of Frankfurt, but essentially by the membership dues and donations of the members of the Senckenberg Research Society, as well as from the liberal grants of generous patrons. However, this financial basis did not prevent the society from regularly running into economic distress and financial embarrassment. The acceptance of the Senckenberg Institutions into the "Königstein Treaty" of the community of the federal states recognized the Senckenberg as an outstanding institution of internationale scientific significance. The Senckenberg did not lose its independance by this operation, however, it gained a regular annual budget. From now on it could endeavour continuous growth of its scientific fields and realize it financially. The exhibits, for which the pre-war conception had been retained even during the restorations after the war, were given a completely new character by Professor Dr. W. Schäfer, zoologist and director between 1961 and 1979. This, however, could only be realized by additional financial help of the government of the Federal Republic. In the framework of the new museum concept within which exhibit hall, graphic designing and displays were mainly treated as an entity, a museum restaurant was also included. In addition, modern environmental problems were was introduced into the exhibits and research programs as early as 1970. At the same time a center for information and documentation of literature in biological sciences was founded. The Senckenberg School for biological technical assistants was opened, the division for anthropology-paleoanthropology was added and an external institute for algology and vegetation history was acquired in the Spessart Mountains. During this period, a small station for hydrobiology and ecology of fresh water areas was opened in north Norway. The annual journal for marine sciences, "Senckenbergiana maritima", and the monograph series "Courier Forschungsinstitut Senckenberg" were first introduced. The marine institute in Wilhelmshaven was expanded both in space and staff and in 1978 a research vessel was commissioned. In 1976 the research institute and natural history museum of the Senckenberg Society were included into the Federal and State government budgets for continuous annual financing.

In the following phase, since 1980, under the directorship of the geologist and paleontologist Professor Dr. W. Ziegler the first task was to consolidate and expand items begun before. Thus, the excavation project of the former oil pit Messel was institutionalized and from now on a regular Senckenberg excavation operation and a research program for Messel fossils was established. In the Kuhwaldstraße in Frankfurt, half a mile away from the main buildings, a former factory building was restored and transformed into a research institute for the Senckenberg geology-paleontology and botany-paleobotany divisions.

In compliance with the increasing importance of marine research, the staff and programs of the marine Institute in Wilhelmshaven are being expanded. In all research fields, cooperation particularly with international researchers and their institutions, is enhanced. The publication series "Courier des Forschungsinstitutes Senckenberg" is being developed as a very important publication organ that appears at short notice and intervals. In the museum, the exhibition halls for insects and crab animals as well as the exhibit on fishes were completely reorganized, the development of a hall for the invertebrate animals was begun and a large exhibit for historical geology and a new hall for a permanent Messel exhibit are being planned.

The educational service, originally established in 1826 in the first Senckenberg Museum was revived in 1981.

Seen as a whole, the Senckenberg institutions serve two main tasks in many ways: teaching and research. Teaching finds its expression in the education of people by the exhibits, the popular publications, and the museum educational service, but also through teaching by many Senckenberg scientists at universities and by scientific publications for students and academics. Senckenberg research culminates in a wealth of scientific tasks in classical natural scientific areas as well as in modern environmental research and in the closely related scientific publications. In summary, the Senckenberg institutions represent a natural history museum and a research institute that has a broad impact, professional diversity and a wealth of scientific results which in its entirety can hardly be found elsewhere.

Editor

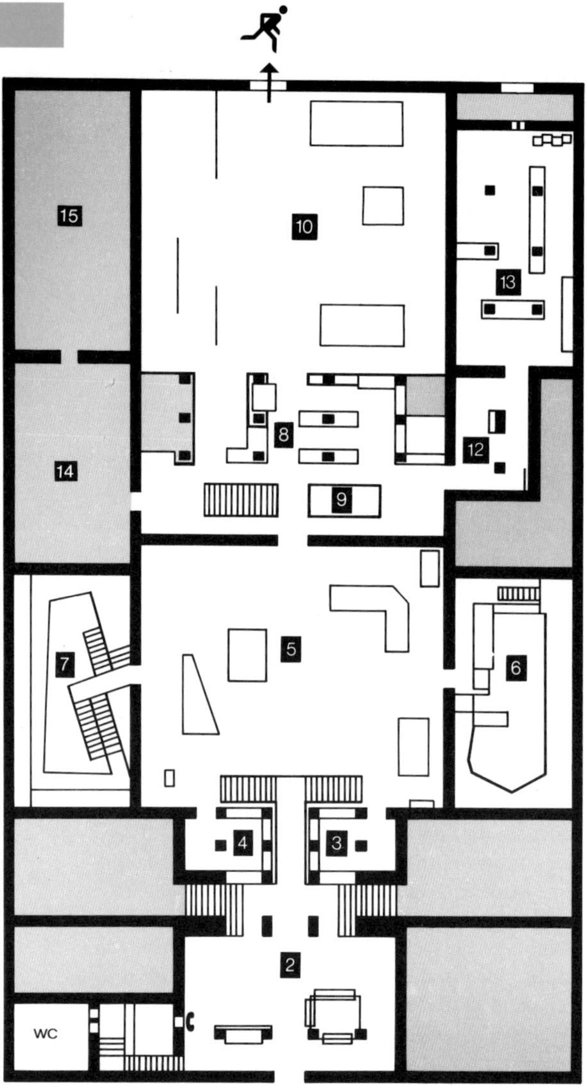
15
10
13
8
12
14
9
7
5
6
4
3
2
WC

Ground Floor: Fossils and Geology

Room 1	Outdoor Exhibits: Giant Ammonite, Fossilized Trunks
Room 2	Entrance Hall: Admission Desk, Telephone, Information Staircase leading to Cloakroom and Toilets Large Slabs from the Early Jurassic Oil Shale of Holzmaden; Crystals
Room 3	Fossils in Mythology (Planned) Temporary Exhibition, Dinosaur Track
Room 4	Ancient Egyptian Animal Symbols, Mummies
Room 5	Central Hall: Dinosaurs
Room 7	Rocks and Minerals, Fossil Invertebrate Animals (Galleries) Temporary exhibition Messel (At Present 2nd Gallery)
Room 8	Fossil Mammals and Vertebrate Fauna from the Mainz Basin (under Construction) Modern Aquatic Mammals, Temporary Exhibition Archaeopteryx, Reptiles (Provisional)
Room 10	Second Hall: Whales and Elephants
Room 12	Biological Affinity of Man and Fossils Wayside to Man
Room 13	Fossils of Man
Room 14	Fossils from Messel (Permanent Exhibition Planned)
Room 15	Earth History (Permanent Exhibition Planned)

1 Outdoor Exhibits (= Room 1)

The Senckenberg Museum, on the street called the Senckenberganlage, is surrounded by buildings belonging to the University and to the Physikalischen Verein. Some large, impressive objects placed outside the main entrance are intended to bring to mind the nature of the wealth of exhibits inside: the purpose is education in natural history. Coming from the north, the visitor encounters a large metallic globe. It shows the position and relative proportions of the six continents. The globe, 3.2 metres in diameter, helps us to understand the dimensions of the planet Earth. The diameter of the Earth at the equator amounts to 12,756 kilometres.
The sun, at the same scale as the terrestrial globe here, would have a diameter of 350 metres, the moon 87 centimetres.
South of the entrance to the museum there is an assembly of fossilized trunks, put together from parts of trunks found in Saxony and Thuringia in deposits of Permian age (Rotliegendes, dating from approximately 250 million years ago). A fountain at the street is built of columns of basalt. These solidified out of volcanic lava some 15 million years ago, during the time when abundant flows of lava were piling up to form the hills of the Westerwald and the Vogelsberg. The form of the basalt columns, hexagonal in most cases, was produced as the lava cooled. To the right of the entrance there is a reproduction in concrete of the shell of a large ammonite. The ammonites, molluscan animals related to squids, were especially abundant among the inhabitants of the seas of Mesozoic age (from 65 to 230 million years ago). They are important to palaeontologists today as a means of dating rocks ("guide fossils") of that Era.
The bust of the Frankfurt physician Johann Christian Senckenberg (1707–1772) placed on the left side of the main entrance serves as a remin-

der of the distinguished traditions of the Senckenberg Nature Research Society and the Senckenberg Museum.

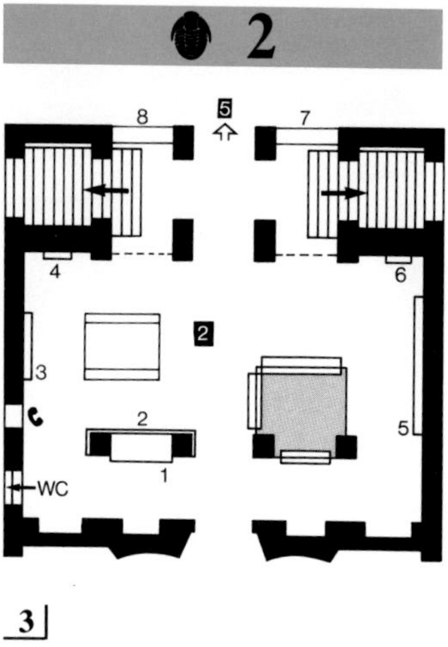

Entrance Hall (Room 2)

When you enter the museum you find on the right the admission desk and on the left the staircase leading down to the cloakroom and toilets. On the walls of the entrance hall are some of the finest specimens of early Jurassic (Liassic) fossils.

Sea-lilies

On the south wall there is a world-famous slab covered with sea-lilies. The sea-lilies *Seirocrinus* rise like a bunch of flowers from a sunken tree-trunk which is covered by an overgrowth of oysters. A similar specimen is exhibited horizontally as it was found on the sea-bottom. The bench around this huge slab invite the visitor to stay for a while and enjoy the beauty of these "flower animals". In spite of their common name they are not flowers but animals which were fixed by a long stem and which fanned out their elaborate, finely feathered crown to filter food particles from the surrounding water. The sea-lilies, together with starfishes, sea-urchins, sea-cucumbers and brittlestars belong to the echinoderms, a highly evolved group of invertebrate animals. All of them have a calcareous skeleton and are occupants of marine environments. There are only relatively few sea-lilies to be found today and they are, in the main, confined to deeper parts of the seas. Sea-lilies were nevertheless abundant in shallow sea areas during past ages and we frequently encounter the dispersed elements of their calcareous skeletons, particularly those from their stems, the trochites, in rocks. Very occasionally the whole skeleton remained in a good state of articulation after the death of the

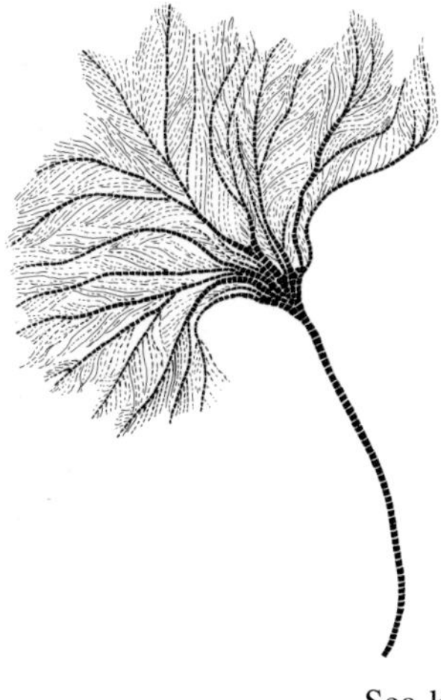

Sea-lily

Marine crocodile (*Steneosaurus bollensis*) with gastric stone animal and slabs with specimens of whole sea-lilies, such as ours, are rarities.

Marine Crocodile 2

The marine crocodile *Steneosaurus*, unlike modern crocodiles which live in fresh water, was an inhabitant of the open sea. It was so much adapted to that environment that its legs, especially the short front legs, were scarcely fit for walking on dry land. The dark oil shales at Holzmaden, in which the marine crocodile was buried, were deposited far from any sea-coast of that time. However, the marine crocodile was certainly obliged to seek the coast, in order to lay eggs, and the flinty stone seen among its bones came from such a place. The stone is enclosed in the dark mass of the contents of the animal's stomach and must therefore have served as a gastric stone which helped to break up food material. Crocodiles cannot chew their prey to any great extent, instead, they swallow it almost whole. Modern crocodiles likewise swallow stones. The fish-lizards, like the marine crocodiles, belong among these reptiles which were no longer at home on dry land.

Fish-lizards

Eurhinosaur (also called "snout lizard"), like any 5
of the fish-lizards of Mesozoic time, was so fully adapted to life in the sea that it was no longer capable of going ashore. The most striking feature of the eurhinosaur is the elongated upper jaw, almost three times as long as the lower. We find comparable jaw development in some modern animals, for example in certain fish (e. g. the sawfish, see Room 208), which seek their food at the bottom and use their long snout to root among the mud. The shape of the eurhinosaur is fish-like, the legs have been transformed into powerful paddles. In the case of the small fish-lizard *Stenopterygius* exhibited here, it is no 6
longer possible to recognize the whole form of

the body. When it sank to the sea bottom, decay of the soft parts was already so far advanced that the individual bony elements of the skull and the tail were already falling apart. The skull bones are now strewn out alongside the skeleton of the torso.

Crystals

1

7–8

Crystals are displayed in three cases in the entrance hall, on the left of the entrance and in the two staircases leading up to the first floor. They are among the largest and finest examples of their kind in the museum. A large part of this impressive collection was presented by one of the great patrons in the museum's history, Arthur von Gwinner. Further minerals are in Room 7.

Forged Fossils

4

As in the case of art works, true and forged fossil objects are often hard to distinguish, even by experts. In many cases it takes comprehensive expertise and technical equipment to identify forgeries clearly.

Two classic examples of forged fossils are on display to demonstrate this: A fossil giant salamander in which skull and tail have been completed by fish bones, and an extant insect pressed upon the surface of a late Jurassic limestone slab thus simulating an alleged age of 130 million years.

◄
Quartz crystals

! 3

Room 3, Fossils in Mythology (Planned) Temporary Exhibitions, Dinosaur Track

Until the completion of the planned exhibition of fossils in mythology, this room will be used for various temporary exhibitions.

The initial "S" from a medieval manuscript in the form of a winged dragon

Fossils in Mythology (Planned)
Without doubt fossils began to fascinate man long ago, and were interpreted mythologically according to the current cosmography. This was the case for Ice Age man and continued up to modern times. Even in the 19th century, after paleontology had been introduced as the science of extinct creatures documented in fossil finds, the strangest, mythologically formed beliefs about fossils were still held in many parts of Europe. In East Asia (China and Japan) such interpretations continued on far into the 20th century. The best known examples are dragons, giants and unicorns. The exhibition shows fossils, fossil traces, and parts of unknown animals which were interpreted mythologically and found their way into popular beliefs. While tales about dragons (including the notions of basilisks, winged dragons, and lindworms) can mostly be traced back to skeletal remains of ichthyosaurs, but also of cave bears and woolly rhinoceroses, legends about giants were based mainly on the skulls of dwarf elephants in the Mediterranean area, and on bones and tusks of mammoths and other elephant-like forms in Central Europe. Tales about unicorns were first occasioned by imported narwhal teeth, but later also by mammoth tusks. Apart from vertebrate finds, the preserved hard parts of some invertebrates also play a part in mythology. For instance, thunderbolts, also called thunderstones, are the vernacular names for the inner skeletons of belemnites, which are classed with the cephalopods. Trochites, which are fossilized skeletal stalks of Mesozoic sea-lilies,

used to be collected under the name "St. Boniface's pennies" or fairy money, and revered for their supposed magical powers. Fossil bivalves and snails were also regarded as miracle cures, e. g., cross-sections of the extinct snail genus *Actaeonella*, which were called "Wirfelsteine" by people in the Alps, and used for their alleged magical power against the gid, a disease of cattle, even into the present century.

Iguanodon Track

The track plate measuring 1 by 2 metres from the Deister sandstone (Lower Cretaceous period, about 130 millon years ago) shows footprints up to 40 cm long made by a dinosaur, the 6 metre high *Iguanodon*. Next to the original stone plate (left) donated to the Museum in 1914, there is a cast (right) showing the imprints made by the feet.

Temporary Exhibition: How does a Zoological Preparator work?

This small exhibition does not deal with a scientific topic, but rather shows some of the work produced by two young zoological preparators during their 2 year's traineeship in the Senckenberg Museum, and testifies to their skill. The specimens are arranged according to the kind of preparation technique employed: taxidermy, small preparations, skeleton mounting, and molding techniques. Each of these methods is illustrated by a series of pictures or models of various stages in the preparation process (e. g., hide, half-finished, mounted, cast specimens, etc.). This small display fulfils a wish often expressed by visitors to learn something about the technique of preparing and exhibiting animals: the famous glimpse behind the scenes.

! 3

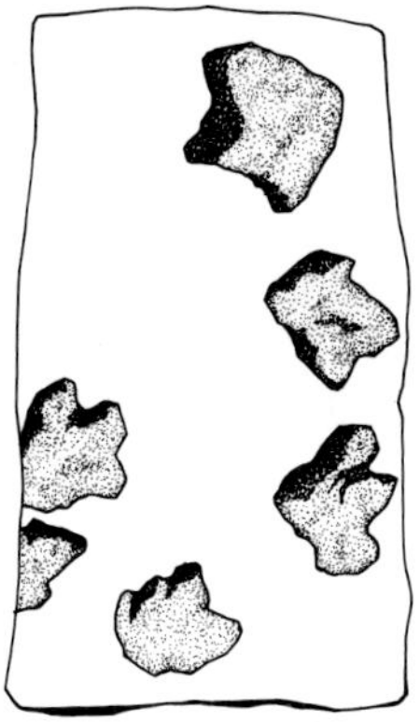

Track of *Iguanodon*

4

Room 4, Ancient Egyptian Animal Symbols

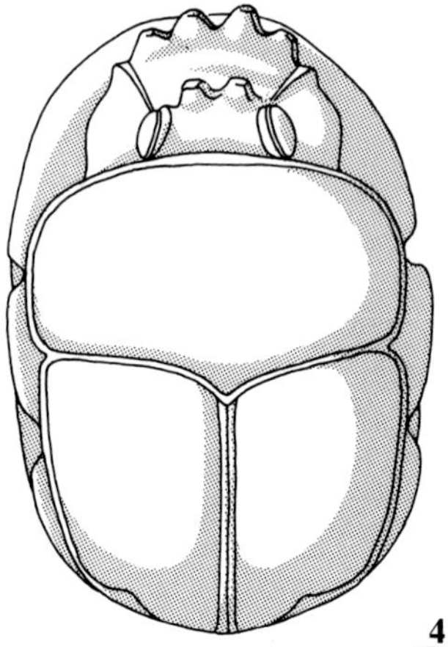

Scarab amulet

It is a matter of course for modern man to think about the animal world in an objective scientific way. In earlier cultures, however, animals played an essential role not only as a source of food but also as creatures with a mythical background. This is exemplified by the display on the cultural history of the animal symbol in Ancient Egypt. Our intellectual logic and objective relation to the animate world would have been completely foreign to the ancient Egyptians; they did not live in a world of objective concepts, but in a world of mythical emblems. It is from this standpoint that the Ancient Egyptians' relationship to many animals has to be understood.

The animal cult played an important role in the Pharaohs' empire; animals served as mediators of superhuman powers and so were worshipped either as bearers of divine revelation or as being
4 themselves divine figures. The falcon, for instance, was believed to be the animal sacred to Horus, the King of the Gods, and was at the same time the symbol of divine kingship. The falcon's wings spreading over the world symbolized the heavens, and were depicted by the "winged sun", in which the sun disk and the two uraei (sacred asps) entwining it formed a further symbol of kingship. The vulture was the symbolic animal both for Upper Egypt, and for its patron goddess, Nekhbet. The snake belonged to the life-giving powers; it was the symbol of the Lower Egyptian goddess of the fields, Uto, and later became the symbol for Lower Egypt. After the two halves of the land had been united, the vulture and the snake were included in the royal regalia, especially in the headdress. The ibis was held to be an incarnation of the god of the moon, Thot, who was also symbolized by the baboon.

5 As an emblem of self-creation, the scarab was equated with the god of creation, Atum, but it

also served as a symbol for the sun god. The bull, Apis, was the holy animal of Ptah, the ancient Egyptian god of creation from Memphis. It bears the sun disk with the uraei between its horns as a sign of its divinity. The cult of the bull was connected with fertility and procreative power, which accounts for the words "Osiris-Apis, mayest thou give life..." on the striding Apis Bull on display, which probably comes from Memphis (XXVIth Dynasty). Besides the bull, the ram was also worshipped as a symbol of procreation in many places in Egypt. The changes in the mythical significance attributed to the various animals can be exemplified by the frog: initially a symbol for the powers of the origin of life, then an animal sacred to Heket, the goddess of birth, and finally the symbol of rebirth. Whereas it was believed at first that the souls of the gods live in certain animals, the animals themselves became divinities as Ancient Egyptian mythology developed further. This was the case in particular for the bull, the crocodile, and the cat. They were holy not only as live creatures, but were worshipped as statuettes and other objects of art as well, and were buried as mummies along with the dead. In the form of amulets, ornaments, statues, rock paintings and murals as well as paintings on sarcophagi, and tablets of stone and clay, Ancient Egyptian "art works" illustrated the beliefs of the time, animal symbols, of course, playing a large role. The Ancient Egyptian picture script, the hieroglyphics, also contained numerous animal symbols (bird shape = soul or life, beetle = origination, lion's head = strength, etc.).
The stars also seemed to be gods to the Egyptians, so they interpreted the stellar constellations and their movements in the heavens mythologically. The reconstruction of a painting on the ceiling in the tomb of King Sethos I (XIXth Dynasty) in the Valley of the Kings (West Thebes) depicts the constellations at the celestial north pole, with the stars shown as red points.

Drafting of an ostracon, the cartoon of ancient times

6

Both humans and holy animals were embalmed and mummified after their death by priests wearing jackal masks as the symbol of Anubis, the god of death. This is shown in the reconstructed mural from the grave of Seenedjem near Deir-el-Medina (West Thebes).
Whereas animal bodies were mostly merely swathed in specially prepared bandages for mummification, human corpses were conserved by an embalming process taking 70 days. The brain and bowels were removed, then the abdominal cavity was filled with aromatic substance, the body treated with natron, anointed, and bandaged. The head was covered with a mask depicting the dead person's face, and the mummy laid in a wooden sarcophagus. The inner organs previously removed were embalmed separately and buried in pitcher-shaped containers called canopic jars. A burial required four canopic jars each sealed with the head of one of the four sons of Horus, as protectors of the inner organs. In addition to mummies of children from Thebes, two limestone canopic jars are on display, one with a falcon head (containing the liver), and one with a dog-faced monkey head (containing the bowels). In accordance with the Ancient Egyptian religion, the sarcophagi were decorated with animal symbols, in this case with the winged sun of Horus, King of the Gods.
The mummies, works of art, and sarcophagi on display were for the most part excavated or bought in Egypt by the explorer and natural scientist Eduard Rüppell, from the Senckenberg Society, between 1817 and 1833, and brought to Frankfurt on the Main.

◀
Mummy of a priest, about 100 BC

1

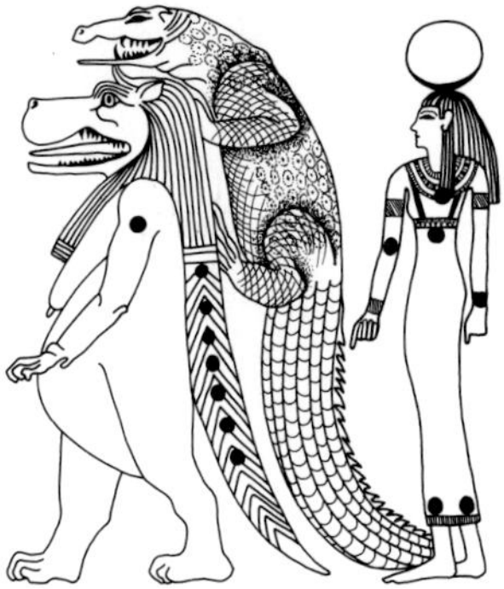

2

Part of a star map on the ceiling in the tomb of Sethos I in the Valley of Kings. The dots characterize the constellation of stars, figurative portrayals of animals and humans are assigned to gods

5

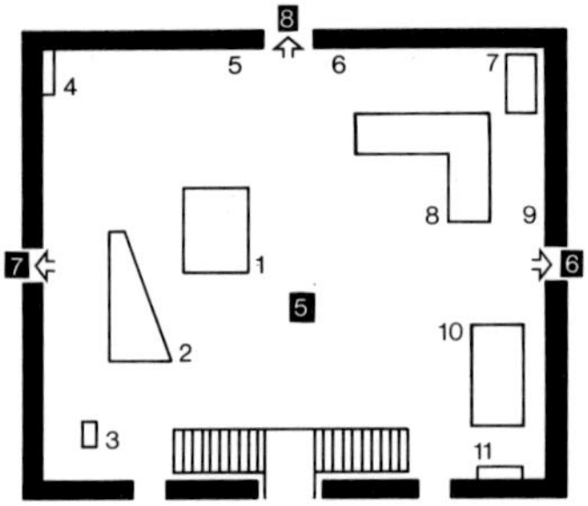

Room 5, Central Hall (Dinosaurs)

The central hall has always been the principal exhibition site in the Senckenberg Museum. A running strip of graphics, set on a dark glass, provides information on the subjects on display, which are principally large reptiles of the Triassic to Cretaceous periods, the dinosaurs, which means, more or less, the "terror lizards". Scientifically seen, there are two groups of large reptiles united under this name: the ornithischian dinosaurs ("bird-hipped", those with a bird-like pelvis), and the saurischian dinosaurs ("lizard-hipped", those with a reptile-like pelvis).

1 **Diplodocus** (Double beam)
The center in this hall is occupied by the skeleton (an original one) of *Diplodocus*, about twenty metres long. It is the only specimen of its kind in Europe. There are five further complete skeletons in North America. Before we descend into the court we should take note of its skull, sixty centimetres long and therefore relatively small. It enclosed a brain whose size was approximately that of a pigeon's egg, and which was minute in relation to the massive bulk of the animal's body, whose weight is estimated at eight to twelve tons and so would compare with the combined weight of two to three elephants. The skeleton indicates that the limbs are massive and heavy, yet the ribs and the vertebrae are quite delicate. The neck vertebrae illustrate a clasp principle of construction. Thus, the weight of the animal was distinctly reduced. Today science suggests that *Diplodocus* lived on dry land, in contrast to a previous assumption that the animal preferred to stay in water in a half-floating mode. Provided with a long neck, the animal could strip off leaves from even very tall trees using its pencil-like teeth. In *Diplodocus* one sees a saurischian pelvis with three major, quite compact radiating elements in its structure.

Skull of *Diplodocus*

Iguanodon (Iguana tooth)
We find an example of an ornithischian pelvis in *Iguanodon.* It has four major radiating elements, is less bulky than that of the saurichian dinosaur and the ischium and pubis bones extend into long, rearward directed projections. *Iguanodon* normally did not run on all four legs but adopted an upright stance. Upright on its hind legs it attained the stately height of six metres. It used its anterior limbs in seizing food, but they could have served also as a weapon of defence, for the thumbs were powerful and armed with an impressive daggerlike claw. The skull of an *Iguanodon* resembles to some extent that of a modern horse, and like the horse, the *Iguanodon* was perhaps an inhabitant of the steppes, feeding on plants. Isolated teeth and bones of this reptile have been found in many places in rocks of Cretaceous age, but whole skeletons have turned up at one place only, near Bernissart in Belgium. The remains of twenty-nine animals were found there. Twenty-three of them are on exhibit in the Natural History Museum in Brussels. We display a cast of one of the best preserved specimens. We occasionally encounter in sandstones tracks made by iguanodons. They reveal impressions made by fore-limbs or the tail. A track of this kind can be seen in Room 3.

2

Movement study of *Iguanodon*

Tyrannosaurus (Tyrant lizard)
The tyrannosaur is one of the largest predatory dinosaurs we know. More or less upright, in its running posture, its 10 metre long body would at its highest point have been 5 metres above ground level. In general form it would not have been greatly different from an *Iguanodon,* but its powerful skull and its very much reduced fore-limbs provide clear evidence to distinguish both. Further, the tyrannosaur fed on large animals and perhaps also on carrion, which we can infer from the powerful teeth in the skull-cast, 1.3 metres long, exhibited here. Carnivorous dinosaurs were

3

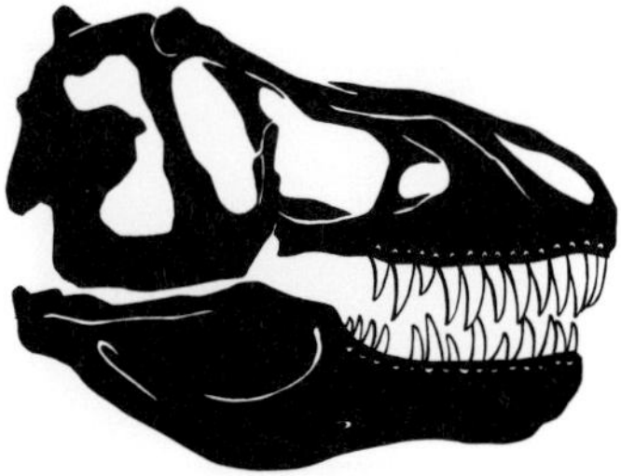

Skull of Tyrannosaurus

extant throughout Mesozoic era. The tyrannosaur, a contemporary of *Triceratops,* lived during Cretaceous time.

4 **Brachiosaurus** (Arm lizard)

One of the largest dinosaurs known to us (and, therefore, possibly the largest land animal for all time) was the brachiosaur. In this case we exhibit only a single foreleg with its shoulderblade. The shape of this gigantic leg is reconstructed with a surrounding brass sculpture. The whole animal twelve metres high, would rise four metres above the ceiling of the hall. Brachiosaur skeletons are known from finds in the Upper Jurassic of East Africa (Tendaguru Mountains). The exhibited limb bones come from the same region. Beds of the same age in North America have also produced brachiosaur skeletons. In North America, the brachiosaurs were alive at the same time as *Diplodocus.*

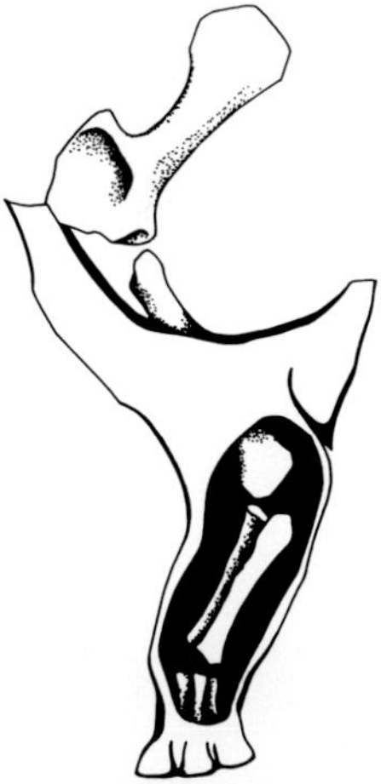

Foreleg of *Brachiosaurus*

5 **Compsognathus** (Pretty jaw)

Compsognathus having an overall length of 60 cm was probably the smallest of all saurischian dinosaurs and illustrates that not all dinosaurs were giants. The selected remains were found in the Solnhofen Limestone at Jachenhausen near Kehlheim (Bavaria) in 1859. *Compsognathus* ran on its hind legs using the long tail as a counterbalance to the upright body. The short forelegs had three fingers armed with claws and were used to grasp and hold the prey.

The cast (display planned) of the only specimen found so far shows great similarity with the *Archaeopteryx* (see Room 8 and 9).

◄

Central hall with *Iguanodon bernissartensis* (left), *Diplodocus longus* (right), and on the wall (background) foreleg and shoulderblade of *Brachiosaurus brancai*

On the wall *Plateosaurus quenstedti,* in front of it the *Triceratops* group with *Triceratops elatus* (skeleton) and *Triceratops prorsus* (skulls)

7 **Plateosaurus** (Flat lizard)
Plateosaurus, the second dinosaur found in Germany, is a primitive saurischian dinosaur of Triassic age, belonging to the group of dinosaurs

out of which *Diplodocus* and *Brachiosaurus* emerged during Jurassic time. The skeleton on display, from the Keuper of Halberstadt, is incomplete, but the presentation in a montage with iron sheeting gives a life-like impression.

Triceratops (Three-horned face)
The *Triceratops* was a particularly heavily built dinosaur with more or less the stature of a large rhinoceros. Its skull bore three long bony horns projecting from the brows and nose. The neck was protected by a thick bony frill, which might have had a regulating effect on the body temperature, cooling or warming up the heavy body. In this way *Triceratops* was well armed to resist attack by its enemies, the tyrannosaur, its contemporary, for example. *Triceratops* fed on plants, which it tore off or rooted up and then boke down with its sharp "turtle beak". The teeth have sharp edges and lie above numerous replacement teeth. The two skulls on display here are originals – and great rarities. One of them is the "heraldic emblem" of the Senckenberg Museum. The group was completed by the cast of a whole skeleton that conveys an impression of the size of these enormous animals.
The triceratopsids were the latest development in the evolution of the ornithischian (bird-like pelvis) dinosaurs, and had their major development in the latest Cretaceous time in North America, where the specimens on display were found. The records there show the existence of forms with only one horn (*Monoclonius*) or with numerous horns (*Styracosaurus*).

Protoceratops (First horned face)
The line of development in the three-horned dinosaurs began with a small form, *Protoceratops,* which had not as yet acquired horns but had a broad frill. From this group we known not only the grown animals, but also all of the young stages and even the eggs. Eggs have been

8

Head and neck region of *Plateosaurus*

9

Protoceratops: reconstruction of skeleton and body outline found still lying together in clutches and in some cases even associated with the bones of baby animals. The protoceratopsid discoveries come from Asia (the Gobi Desert).

Stegosaurus stenops

10 **Stegosaurus** (Roofed lizard)
Stegosaurus is a 6 metres long ornithischian with bird-like pelvis. The original of the exhibited cast came from the same locality which had yielded our *Diplodocus*. The most characteristic features are the bony plates along the backbone and 2 pairs of strong spines near the end of the tail. These plates were well equipped with blood vessels the impressions of which are preserved; therefore they may not only have served as a

protectional device, as widely believed, but also as regulatory organs for the body temperature. The arrangement of their backbone plates was alternating rather than in pairs, two equally sized plates suggesting a symmetrical arrangement in pairs have never been found.

Ophthalmosaurus (Eye lizard)

11

The fish-lizard *Ophthalmosaurus* supplies a link into the display in Room 6. This animal was a member of that group of reptiles which, like the whales among mammals later in time, became adapted to life in the sea. Like the plesiosaur in Room 6, this fish-lizard, dating from Late Jurassic time, comes from southern England.

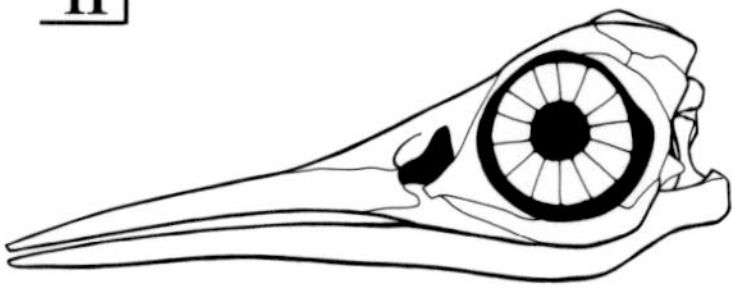

Skull of eye lizard

Dimetrodon (Two-sized tooth)

6

Just as the fish-lizard has awakened our anticipation of the exhibits in Room 6 (saurians with aquatic mode of life), *Dimetrodon* is to call attention to the fossil mammals. *Dimetrodon* is not a dinosaur, but belongs to the original group of mammal-like reptiles, the landdwelling pelycosaurs from Late Carboniferous to Early Triassic. The most striking feature is the array of extraordinarily long spiky projections from the spine. These supported a high dorsal ridge the function of which is not fully understood. It may have regulated the body heat. The teeth show that *Dimetrodon* was a predator.

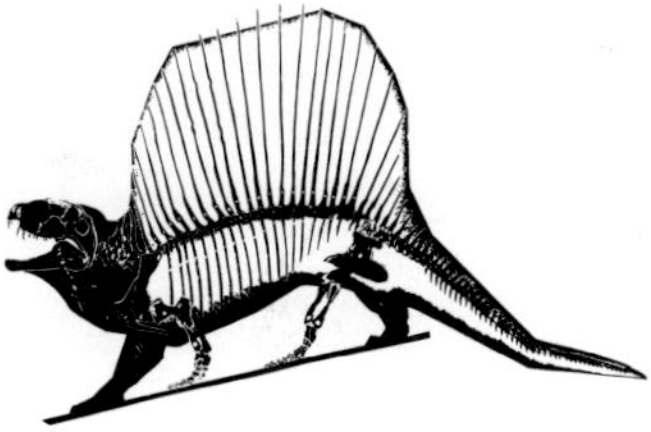

Dimetrodon (reconstruction)

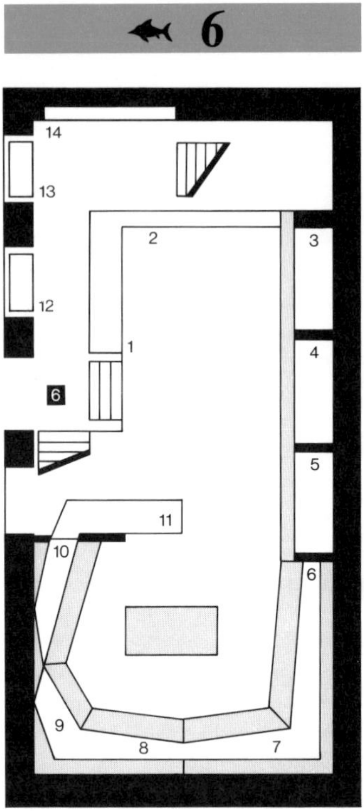

Room 6, Fossil Vertebrates Adapted to Aquatic Life

This room, leading off from the northern side of the central hall, houses a display of fossilized examples of various vertebrates which were well adapted to life in water. At the lower level we meet several representatives of the Mesozoic swimming reptiles. The upper level is principally concerned with fish and the examples range from Paleozoic era to Recent.

In both the past and the present time, the sea has provided niches and nutriment for a wide variety of animals. It is no great surprise to find that in addition to those groups of animals which have, from the times of their origins, lived in the sea, there are many examples of land animals having returned to life in the sea. Examples among present day mammals would include the whales (Room 10), the sea-cows and the seals (Room 8). During the Mesozoic era, the time when reptiles were predominant, the inhabitants of the seas included forms derived from several different reptilian groups:

marine crocodiles (among the large Archosauria), the mosasaurs (among the Lepidosauria, to which the lizards and snakes belong), marine turtles, and whole groups like the ichtyosaurs, the mesosaurs, the sauropterygians, and the placodonts. Left of the entrance door, a geologic time scale is provided as well as a graphic reviewing all these groups, each of which achieved, quite independently of the others, adaption to life in the sea.

2 *Dimetrodon* will be soon remounted in the central hall (Room 5) and has, therefore, been mentioned in that section (page 31).

An example of especially good adaptation to aquatic life is provided by the sauropterygians. They include the nothosaurs of Triassic age and the plesiosaurs which lived from Triassic to Cre-

Plesiosaur *Peloneustes philarchos*

taceous times. All of them have a sturdy body, powerful and obviously mobile paddles and a relatively long neck.

Plesiosauria 1

In the plesiosaur *Peloneustes,* the torso is rigidified by the powerful ribs, large bony plates in the pectoral and pelvis girdles and the interserted belly ribs, thus approaching the form of a turtle. These animals must have moved like turtles, making powerful downstrokes with their long tipped wing-like paddles ("flying under water"). This agility was reinforced by the long neck, which provided the capacity for sudden movement, and thus the large jaws with strong teeth could seize prey at flash speed. *Peloneustes* lived during Late Jurassic time. The skeleton on dis-

Movement study of plesiosaur

play comes from Peterborough in England. No complete skeletons were found there. It is necessary to assemble a skeleton from isolated bones or portions of skeleton which may originally have belonged to several different individual animals. In Cretaceous time a trend in plesiosaur evolution produced longer and longer necks. The peak development of this kind is seen in the late Cretaceous form *Elasmosaurus* whose total length of 13 metres included a trunk only 3 metres long but a neck extending to 7 metres (with the tail and the 0.6 metre long head con-
6 tributing the remaining 3 metres). *Cryptocleidus* ("the hidden clavicle"), is a plesiosaur too, similar in structure to *Peloneustes* but with a heavier trunk and a smaller head. It was found in the same area that produced the *Peloneustes* skeleton and also the fish-lizard *Ophthalmosaurus* on display in the central hall.

11 **Nothosauria**

A second group of sauropterygians is that of the nothosaurs. They are on exhibit in a flat island case. The nothosaurs were extant during the Triassic in the shallow sea which covered central Europe at that time. Remnants of them are fairly common in the Muschelkalk. Their capabilities as swimmers did not match those of the plesiosaurs and they would have been quite capable of travelling freely on dry land. The sturdy trunk, the only slightly altered limbs and the long neck supporting a flat, elongated head, are all clearly to be seen in the specimen on display here. By comparison with the plesiosaurs, the nothosaurs were small. Nevertheless, some species reached a size of 2.5 metres or more.

Placodontia

The placodonts were contemporaries of the nothosaurs and lived, like these, in the shallow Triassic seas of Central Europe. They were only poorly adapted to aquatic life, their bodies were

Skeleton and image-reconstruction of placodont
Placodus gigas

powerfully developed, and the belly ribs and bony dermal plates contribute to an appearance which resembles that of the turtles. This resemblance is most distinct in the specimen of *Henodus* on exhibit. It comes from Keuper beds in the Tübingen neighbourhood (the original is in the Tübingen Museum). 9

An original specimen here, and an especially valuable one, is the *Placodus*. Apart from the skeleton on display, only isolated bones, skulls and teeth of placodonts have come to light. Teeth are fairly common, because the animal frequently 10

Reconstruction of *Placodus*

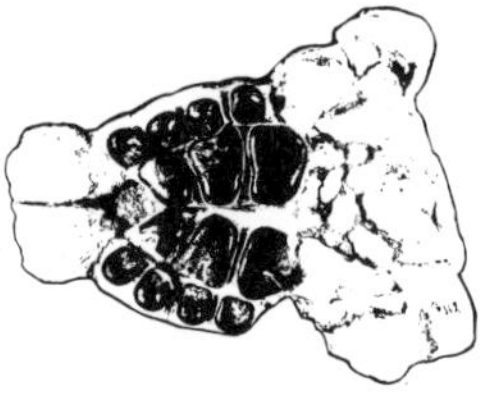

Lower jaw of *Placodus*

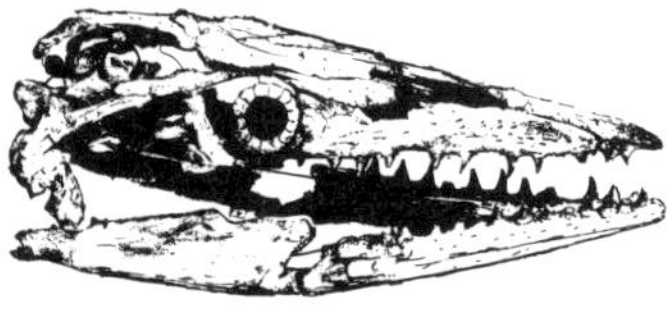

Skull of mosasaur

replaced its dentition. Placodont dentition was a robust one and enabled the animals to feed on shelly animals like mollusks. They captured them or tore them loose with their incisor teeth, and used their flat-crowned cheek and palatal teeth to crack them open. Numerous other inhabitants of water have a "nutcracker" dentition of this character, e. g. *Lepidotes* (Room 9). The story of the discovery of the only known placodont skeleton is an exciting one. It was first recognized only after the enclosing limestone had been broken down. The individual fragments were patiently sought among stockpiled limestone fragments, were stuck together, and the skeleton was then prepared out from one side. The other side was then prepared and finally the whole specimen was free and ready for mouting.

7 **Mosasauria**

The mosasaurs, represented here by a *Plateocar-*
8 *pus* skull and a *Tylosaurus* skeleton from North America, were first discovered near Maastricht in Belgium. The derivation of the mosasaurs deserves a comment: they are closely related to the monitor lizards. During late Cretaceous time, when the fish-lizards and plesiosaurs were already on the point of extinction, this marine group evolved out of terrestrial lizards into marine animals, reaching a length of twelve metres. Predatory, capable swimmers, they were in some respects similar to the fish-lizards and it could be said that they exploited the period of time intervening between the extinction of the fish-lizards and the emergence of the whales, early in Tertiary time.

Marine Crocodiles

One already meets a marine crocodile from the Holzmaden oil shales in the entrance hall. In this room, on the east side, we encounter another, a
7 representative of the genus *Metriorhynchus*. It comes from the Upper Jurassic of England, like

the plesiosaurs, and like these again it is here mounted free. The marine crocodiles represent a mode of swimming quite different from that practised by the other reptiles in the room. The powerfully developed hind legs and tail provided the drive. The fore-limbs were more or less dwarfed.

Fish lizard (*Stenopterygius crassicostatus*) with unborn young animal

Fish-lizards (Ichthyosauria)
The ichthyosaurs on display all belonging to the genus *Stenopterygius* – the previous name was *Ichthyosaurus* – come from the Lower Jurassic (Liassic) oil shales of Holzmaden. These rocks are also the source of the plates covering the walls of the room. This.kind of rock would have originated as an accumulation of fine mud at the bottom of the sea. The bodies of marine animals,

sinking to the bottom, would be slowly buried there in an environment low in oxygen and the corpses would therefore suffer relatively little disruption on decay. We therefore find them in the rocks now as well preserved fossils. Some of the vertebrate skeletons have all of their bones still in the same relative positions as in life. In exceptional cases, soft parts of the body have survived, as shown here by the skin of the ichthyosaur. The fact that preservation is not in every case so spectacularly good as indicated in
4 the display by the example of an ichthyosaur skeleton whose individual bones have fallen apart
5 and by the specimen shown still embedded in the rock. In this latter case the long patient work required to "exhume" the skeleton would scarcely be justified. This specimen is on display to present the first moment of discovery and to give an idea of the wealth of fossils contained in these rocks.

The extreme example of reptilian adaptation to aquatic life is provided by the ichthyosaurs. The clearest case for this adaptation on display here is
3 the specimen of *Stenopterygius* in which an outline of the body is preserved. One observes the form of the body and also the fins, not all of which were supported by bones (dorsal fin, upper fluke of the tail fin). The fish-lizards were fast, active swimmers comparable to the modern sharks and dolphins. The fins derived from limbs, together with the dorsal fin, acted as stabilisers and served to steer the streamlined body whose drive was provided by the tail. There was little possibility of moving the head relative to the trunk, because the series of neck vertebrae is short and stiff. It is possible that the strengthening of the lower fluke of the tail fin (provided by the terminal part of the backbone) helped the animal to emerge rapidly to the surface of the water where it could breath air from the atmosphere into its lungs.

6

14

There is only one three-dimensional exhibit of fish-lizards in the Museum: *Ophthalmosaurus* in the central hall (Room 5).
On the way to the upper floor gallery one passes two fish-lizards mounted on the wall. The lower one is a mother with a young animal in her body. The black mass below represents the stomach contents. Reptiles lay eggs, thus marine reptiles were obliged to return to land, or as an exception to develop eggs within the body and give birth to live young. This strategy is called ovovivipary. Fish-lizards, being incapable of going ashore, certainly brought their young into the world alive. Several fish-lizard mothers with embryos preserved within their bodies were found at Holzmaden. One of the best of these specimens is on exhibit here.
In the upper floor gallery there are various show-cases with fossil fishes. This display is under construction and will be completed in the future by adding fossil amphibians.

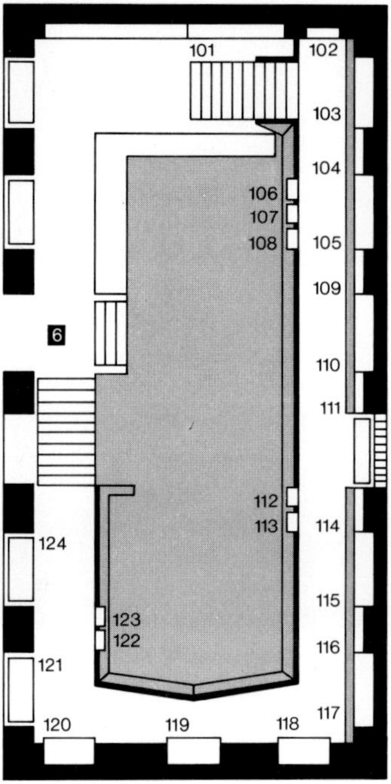

Armoured Jawless Fishes (Ostracoderms) 103

Fish-like vertebrates were found in rocks as old as the Silurian. These primitive armoured fishes had a bony external skeleton like the placoderms, but there is no sign of bony jaws. Because of this latter characteristics, they are classified, along with the present day cyclostomes (lampreys and hagfishes) as the jawless group, the agnaths. Two representatives of the primitive armoured fishes are on exhibit: *Pteraspis* from the Early Devonian Siegenian Beds and *Drepanaspis* from the well-known Hunsrück-Slate also of Lower Devonian age. The flattened *Drepanaspis* is not only exhibited as fossil, but also in a diorama showing its environment and mode of life (see case 118).

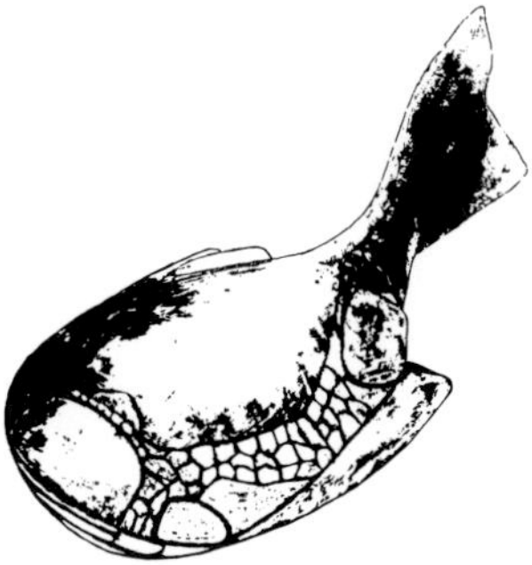

Reconstruction of an ostracoderm

Cartilaginous Fishes (Chondrichtyans) 105

The cartilaginous fishes may have derived from placoderms (not on display) which evolved in the Paleozoic since Silurian times. Since Mesozoic

Armoured jawless fish *Drepanaspis gemündensis*

era cartilaginous fishes have been represented by forms which still live in the present day seas: sharks, rays and chimaeras. The most beautifully preserved specimen of the fossil fresh water shark *Orthacanthus senckenbergianus* is on display in Room 208 on the second floor.

Three small cases display important fossilized elements of the fish skeleton.

106 Teeth are often the only preserved remains of fish. This goes especially for cartilaginous fish, whose teeth fall out after use and are replaced by new ones. The teeth are sources of information in regard to family and species classification but also on feeding habits. The internal and external skeleton of fish is made of bone and cartilage. Bone is much more likely to be fossilized than cartilage. Even in fossilized skeletons, it is quite easy to distinguish skull, vertebral columns and fins. In bony fishes the skull and the body surface give evidence of advanced ossification of the skin. The scales of the cartilaginous fishes are developed as dermal denticles (placoid scales) having a long basal plate and a tooth-like upper part. Early bony fishes mostly have thick shiny "enamel scales". In modern bony fishes the scales are thin, transparent little plates whose composition is bony.

Chondrosteans (Cartilaginous ganoids) **110**

The cartilaginous ganoids (Chondrostei) are the most ancient forms of the bony fishes. Their internal skeleton is still largely cartilaginous, their skulls, on the other hand, are strongly ossified. The body is covered by shiny "enamel scales" (ganoid scales). From Carboniferous to Triassic time there was a great variety of cartilaginous ganoids. Today, the sturgeon-like fishes (Acipenseriformes) are the only surviving representatives of this group.

Two further cases show ear-bones (otoliths) and concretions. The ear-bones of fish are calcareous formations which lie in the otic capsules of these animals. They assist in the sense of balance. They are often the only remaining parts of fossil fish and they are a source of information on the original distribution of the fish, to which, judged by form and fine structure, the ear bones belonged. Concretions are growths within porous sediments which may enclose fossils. The precipitation process which produced the concretions may have been stimulated by some of the substances released during decay of the buried animal.

107

108

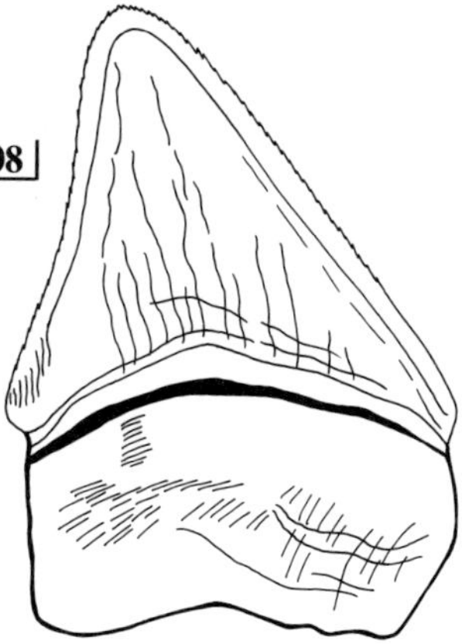

Tooth of a fossil shark

112

113

Different ear-bones of fish

6

Holosteans (Bony ganoids)
The bony ganoids (Holostei) represent a transi-
114 tional stage with many forms between the car-
tilaginous ganoids and the true bony fishes. The internal skeleton was ossified in part, the skull capsule was surrounded by a strong bony armour. The bony ganoids emerged toward the end of the Palaeozoic (Permian) and were abundant during Jurassic and Cretaceous times. Their only present day representatives are the gars (*Lepisosteus*) and the American bowfin (*Amia*).
Mode of life and environmental habitat can be inferred from the form of the body, the develop-
115 ment of the fins and the dentition, even in extinct
fish. Fast-swimming inhabitants of the open sea like *Pachycormus* have a long slim body and a large, deeply divided tail fin. Dwellers in quieter water often have a high body and a rudder-like tailfin like *Dapedium*. Predatory fish have sharp teeth, feeders on hard-shelled animals have
116 batteries of rounded teeth which may cover the
whole roof and floor of the mouth.

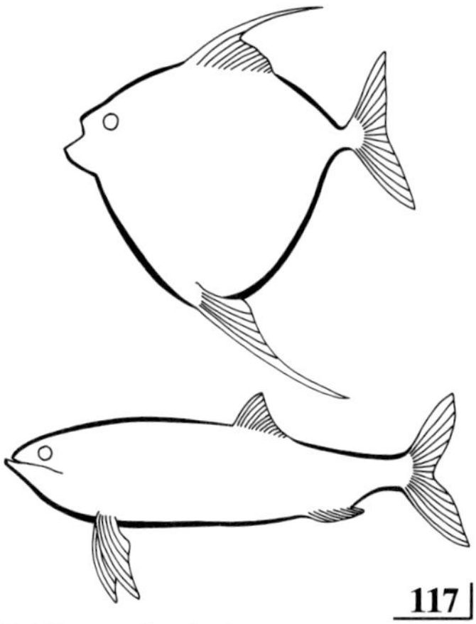

Different body forms

Teleostians (True bony fishes)
The true bony fishes (Teleostei) first evolved at a time close to the change from Jurassic to Cretaceous times. They are the most varied group of all the fish-like vertebrates of the present day. Their skeleton is normally fully ossified. One of the most primitive forms was the herring-like *Leptolepis* (see Room 9, case 6) whose scales were still covered with shiny, enamelled ganoin. The true bony fishes were already the dominant fish group by Early Tertiary time.
The only true bony fish displayed in this room is on a rock slab from the Middle Eocene beds from Monte Bolca/Italy showing two sides of a speci-
117 men of *Gasteronemus*. The diversity of present
day and fossil bony fishes is on display in Room 207 on the second floor.

Lung fishes and Coelacanths

The lung fishes, Dipnoi, emerged as a distinct group during Palaeozoic era (Early Devonian). The early lung fishes, like the present day ones, had a lunglike modification of the swim-bladder and were able to take their oxygen from the air. The fossil coelacanths, Crossopterygii, too, had lungs as well as gills, and were able to use both. Since these fishes have many details of anatomy in common with the primitive amphibia, they are regarded as the root-stock of the terrestrial vertebrates. The coelacanths were already extant during the Devonian, i. e. in the Palaeozoic. The only form surviving at the present day is *Latimeria*. Present-day lung fishes and *Latimeria* are on display in Room 207. A detailed phylogenetic tree of the fish-like vertrebrates and fishes is found in Room 208, case 3 on the 2nd floor. 119

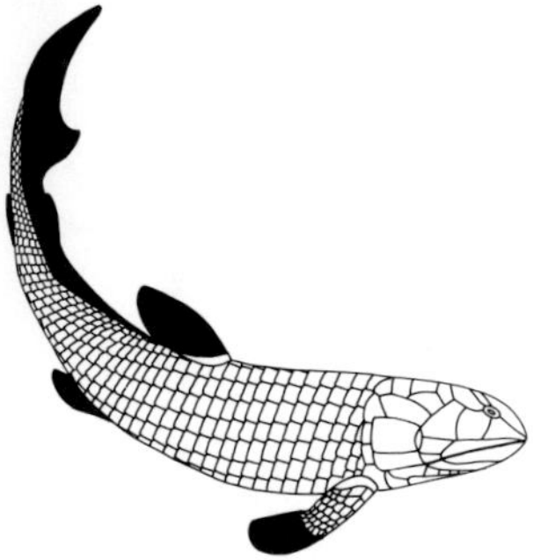

Reconstruction of a fossil lung fish

Palaeozoic fishes are shown on the east walls in reconstructions of their appearance in life (dioramas). One group comprises armoured fishes from the Early Devonian, the other shows the primitive coelacanth *Eusthenopteron*. Early in Carboniferous time, the first terrestrial tetrapod vertebrates evolved from coelacanths like these. 118 120

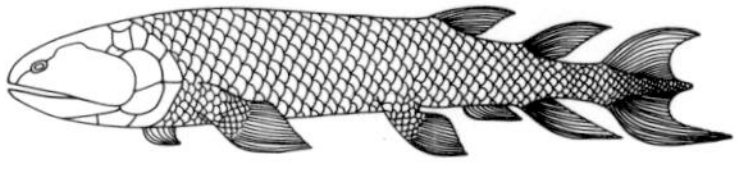

Eusthenopteron

The last two cases show fossilized eggs. A fossil ostrich egg found on the island of Lanzarote is proof of the fact at the Canary Inslands were once joined to the African continent as the African ostrich is a ratite, a cursorial bird, which can neither swim nor fly. Reptilian eggs are rare fossils. However, some have been found (abundant shell fragments and a few complete eggs) in Cretaceous deposits near Aix-en-Provence. Clutches of protoceratopsid eggs even with embryos are known from the Gobi desert. 123 122

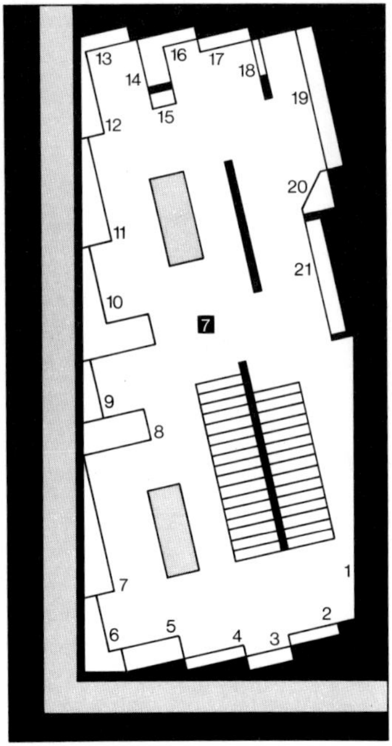

Room 7, Rocks and Minerals, Fossil Invertebrate Animals, Temporary Messel Exhibition

The three levels in this side room which is an avantgardistic architectural design are accessible from the central hall and provide a survey of the component materials of the earth's crust (rocks and minerals on display in the basement), and a review of the wide variety of fossils as evidence of earlier invertebrate life (galleries). In the 2nd gallery some excellent examples of fossil records from the world famous fossil site of the Eocene Messel oil shale are on display in a temporary exhibition. We start our round trip in the basement.

Rocks and Earth's Crust

The solid earth's crust is relatively thin in relation to the 12,576 km equatorial diameter of the globe. At depths of 30–50 km the mantle begins and the core is calculated to begin at a depth of 2900 km. Between the mantle and the core there is a transitional layer (1200–2900 km range). The form of the earth's surface is affected by influences of two kinds: by the internal energies of the earth, which cause crustal movements, and by effects imposed from outside, from the atmosphere, by solar radiation or as the results of the major cycling of the waters on the earth's surface. The various rocks of the crust undergo a perma-
3 nent circular reshaping over millions of years: they may be destroyed, newly formed or transformed into one another. According to their genesis, composition and texture, rocks are classified into three groups. The show-cases for each of these groups also contain those minerals characteristic of the respective rocks.

View of Room 7, basement: rocks and earth's crust ▶

7

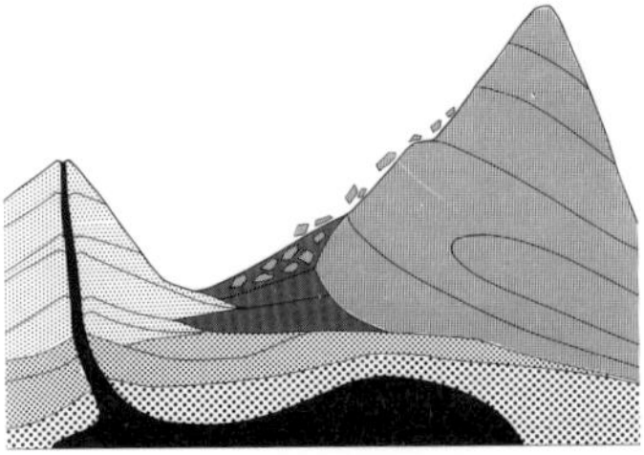

Origin of sediments

Sedimentary Rocks
Destructive forces of weathering processes and
2 mechanical forces of water, wind and ice crumble and erode the surface of rocks. The debris is transported away and deposited as sediments at different places. This reshaping process is slow
4 but can be observed today, too. Examples of sedimentary rocks are displayed: a sequence of sand, pebbles, limestone, and coal. The layers of deposited sediment contain remains of prehistoric animals and plant life. Thus, the sedimentary rocks yield the most important records of earth history. The sedimentary block on display – the Cheops-Pyramid in ancient Egypt was built from rock material like this, shows densely packed coin-sized flat shells of nummulites, protozoans that lived by the billions in Late Cretaceous and Early Tertiary seas.

Igneous Rocks
These rocks are derived directly from hot molten magma. In rising to the earth's surface, they often pour out in sudden volcanic eruptions. Some of these volcanic rocks solidify, however, shortly before they reach the surface, an example
6 of which are the regularly shaped basalt columns. Lava flows on the surface, volcanic ashes, later hardening into tuffs, and volcanic bombs thrown out at the same time are the visible forms of volcanism that takes place at the earth's surface.
7 The skull of a Pueblo Indian gives evidence of the fearful nature of a volcanic eruption.
Punical rock is thrown out in the initial gaseous phase of a volcanic eruption; as it contains numerous pores now filled with air, it floats on
8 water. Quite different is the obsidian, a volcanic glass, derived from gas-poor magma cooling down rapidly.
In dealing with magmatic rocks one contrasts volcanism with plutonism, by which deep-seated magma stocks give rise to plutonic rocks and dyke rocks deriving from them. Diorite and gra-
9 nite are plutonic rocks.

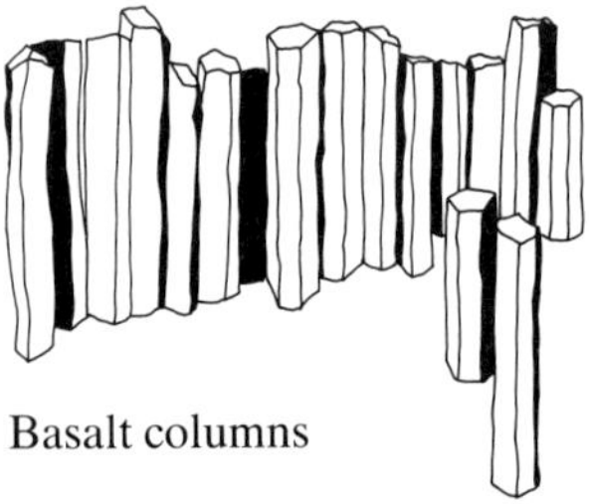

Basalt columns

Metamorphic Rocks

Metamorphic rocks are formed, like plutonic rocks, in slow processes within the earth's crust. They do not develop from magma but are formed by the action of heat and pressure on existing rocks. Rocks become buried in great depths by movement of the crust and will then be dissolved under high temperature and pressure. In ascending they cool and re-solidify with a different
texture. Typical metamorphic rocks are gneiss 10
and marble. 11

Meteorites and Moonrocks

Nearby material of extraterrestrial origin, "cosmic rocks" are on display, which fall on the
earth's surface as meteorites. The large iron 12
meteorite from South Africa weights 328 kg, and
compared with it the fragments of moonrocks 13
look especially tiny – yet everyone should reflect on the huge investments of money and skill that were necessary in order to bring this scientifically so eloquent material to our earth. Thanks to the research work done on such specimens we have now not only learned something of the history of the moon, but we also made the discovery that the moonrocks are composed almost exclusively of minerals which are already known on earth. Examples of such minerals are on display beside the moonrocks.

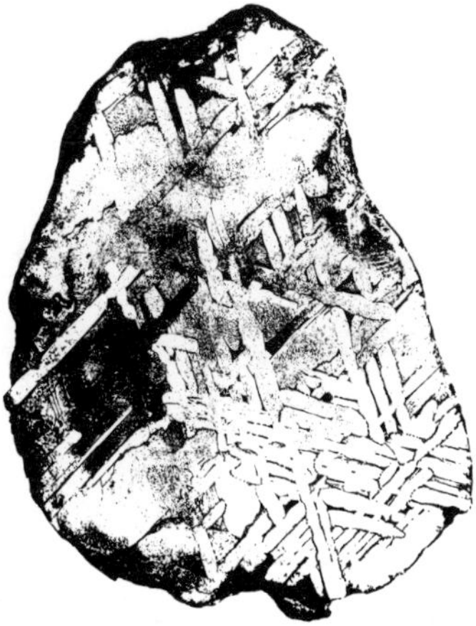

Iron meteorite

Minerals

Minerals are the natural building materials of the earth's crust. There are approximately 2000 different kinds of minerals, of which perhaps only about 200 are of common occurrence. Particular attention is paid to minerals when they are well crystallized. Two showcases, not yet completed, will provide information on particular characteristics of minerals such as colour, size, crystal form and their capacity for refracting light. The last four cases offer a review of minerals. The system of classification used is based on chemical composition. We distinguish 9 classes of minerals.

18 Of special note is the 1st class of the native elements, such as some precious metals, like gold, silver or platinum which occur in their elementary (native) form in nature. Carbon, which is present in many compounds, can nevertheless occur as an element in nature in the form of graphite or of diamonds.

19 In the 2nd class: sulphides and 3rd class: halides there are some economically important minerals. Examples are the gold gleaming pyrite and the rock salt. Many important minerals, including several kinds of ore, belong in the 4th class of oxides and hydroxides. The most abundant mineral of all, quartz – beautifully coloured variants of which are known – is given a case to itself. In the 5th class of the nitrates, borates and carbonates, the most common mineral is without doubt calcite and the one with the greatest range of delicate colours is fluorite. A wellknown mineral in the 6th class of sulphates is gypsum. Some of the minerals in the 7th class of the phosphates are important as sources of fertilizer.

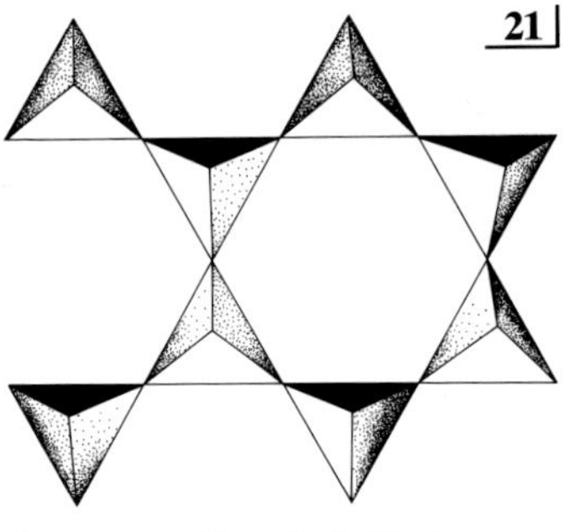

Structure of banded silicate

21 The 8th class of the silicates takes up more space in the display because the minerals following next in abundance to quartz, in fact several of the most important rock-forming minerals, belong here. It is necessary to mention only garnet, augite and hornblende, micas and feldspars. Indeed, the various feldspars, taken altogether, make up a larger proportion of the earth's crust than quartz does. In the case of the silicates their crystal structures are shown, too. This helps one to understand the way in which the individual atoms are associated with one another in three-dimensional lattices. Among silicates, the basic unit in the lattice structure is the SiO_4 tetrahedron (4 oxygen atoms lie at the apices of a tetrahedron which encloses a silicon atom). The arrangement of the SiO_4 tetrahedra in space determines many of the properties of a silicate

mineral – and in consequence the technical possibilities of exploiting the mineral economically.
The last of these nine classes comprises the organic compounds. They are, however, mixtures of minerals in most cases (amber is an example) and are therefore, strictly speaking, rocks.

Next to the stairway up to the 1st gallery, the foundation of the Museum Building erected 1904 is exposed: a layer of coarse pleistocene pebbles, 1 metre thick, is overbuilt by a mortared wall of roughcut sandstone from the Upper Triassic Keuper. **22**

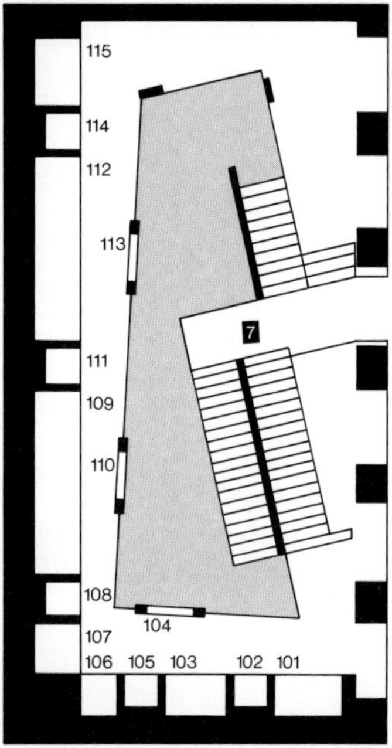

Invertebrate Fossils
Both galleries of Room 7 are reserved for invertebrate fossils. The 2nd gallery however, still houses the temporary exhibition of Messel fossils.

Sponges (Porifera) **104**
The sponges are among the simplest of the multicellular animals, as they have no specialized tissues or organs. They are generally sack-like in form, with no set external shape. With the help of specialized cells, called choanocytes, they contrive to drive streams of water inward through their porous walls, through their central cavity and upward and out at the open top. The sponges' bodies are supported by delicate little calcareous or siliceous needles which may be intergrown with one another to form a compact skeleton. Sponge skeletons and needles have been found in marine deposits ranging in age from Early Palaeozoic time onward. Sponges are known to occur in colonies and can in fact produce small reefs (there are some examples of sponge plugs in the Upper Jurassic of Baden-Württemberg).

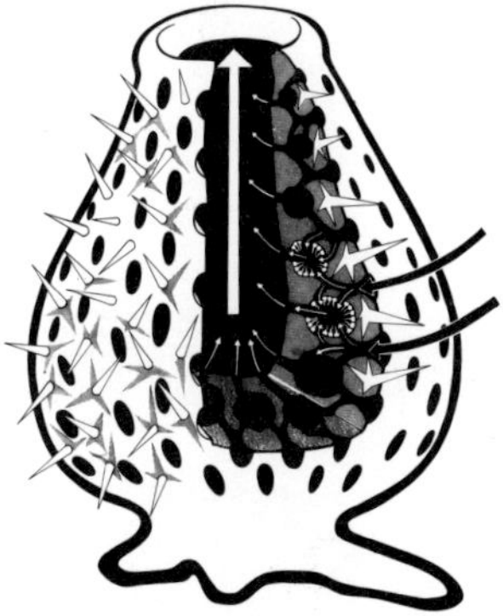

Diagram of a sponge

7

101–103
105–107

Cnidarians (Cnidaria)
In the cnidarians there are two quite distinct life forms: first the medusae, the jellyfishes which travel freely through the sea and secondly the attached, immobile polyps. The cnidarians include the corals, which are relatives of the sea-anemones, and which through a long range of geological time have been one of the chief builders of reefs. The polyps of corals are crowned by a wreath of tentacles. From their basal parts they secrete a cup-like investment of hard calcium carbonate which surrounds them as they grow up. The secreted mass of carbonate is not solid throughout, as the structure of the polyp is reflected within it in radial lamellae (the septa) in vesicular growths and in horizontal tabular plates (the "floors"). Reef corals live mainly in colonies built up by innumerable polyps which reproduce

Bryozoan case

by budding. The coral reefs of the past were constructed by kinds of corals which are now extinct and which were different from the corals of today. The thick masses of limestone on Gothland (Silurian) or in the Eifel, the Bergisches Land and the Sauerland (Devonian) all bear witness to their activities. Many quarries are now important sources of marble, cement, and lime.

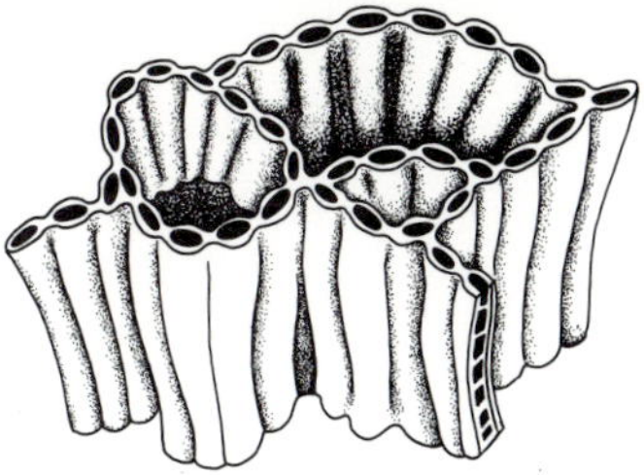

Fossil tabulate coral

Moss Animals (Bryozoa) **108**

Bryozoans or moss-like animals regularly produce colonial growths which often encrust either the sea-bottom or the shells of marine animals living on the sea-bottom. The tiny individuals, each with a wreath of tentacles, are not polyps, however. They are more worm-like and their internal structures resemble those of brachiopods.

Lamp Shells (Brachiopoda) **109**

Brachiopods have a shell comprising two valves which are made of calcium carbonate or, more

Brachiopod case

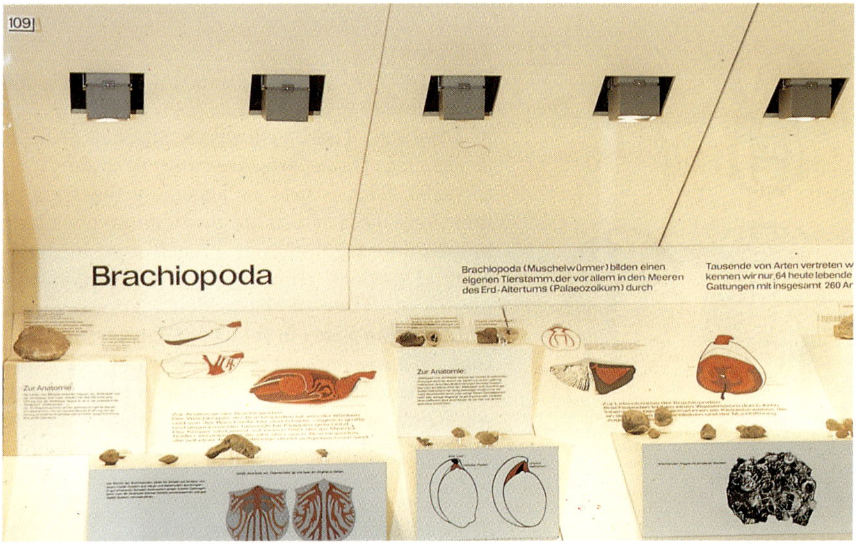

rarely, horn. They are not related to mollusks, as one can immediately perceive by comparing the symmetry of the shells of these two different groups. If one opens the valves of a brachiopod one sees a large, looped or spirally wound pair of "arms" bearing numerous cirri. These cirri are mobile and produce currents of water which allow the animal to obtain the fine plankton on which it feeds. The brachiopod is anchored to the bottom mostly by a fleshy stalk and the animal can move by swinging around its point of attachment. Brachiopods have no great importance among the populations of the present-day seas; but during the Palaeozoic and to a lesser extent in the Mesozoic era they produced a huge variety of forms, and some beds of rock dating from those times are almost entirely composed of banked up
110 masses of brachiopod shells.

The arthropods, the jointed-leg animals, gave rise in Palaeozoic era to mainly marine forms which are completely different from present-day representatives of the group. The following types are important as fossils.

112 **Trilobites**

Trilobites were so abundant during the Palaeozoic that this Era has been called the Age of Trilobites. These arthropods, their carapace is divided into three parts, are closer to spiders than to crabs. This is best indicated by the form of their long limbs which are divided into a walking branch and a gill-bearing branch used in swimming. However, trilobite limbs are only rarely preserved. Some of the evidence known comes from roofing slates in the Devonian of the Hunsrück. Bizarre members of the group were the horned trilobites (see the plexiglass cubes) which were only with great difficulty prepared out of the rock and whose delicate form now requires

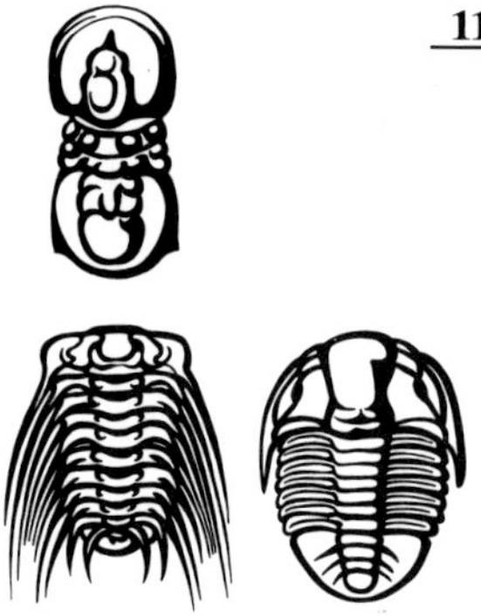

Different types of trilobites

Trilobite (*Ameura major*) ▶

special protection. A slab from the Cambrian of China is covered with the disarticulated elements of trilobite carapaces. Slabs of this kind are traded as objects of art in China.

113 **Merostomata**

The horseshoe crab (*Limulus*; see Room 204 on the 2nd floor) is in fact not a crab but is, like the scorpiones, related to the arachnids. Its larvae resemble trilobite larvae. Horseshoe crabs have changed very little during the course of earth history and have survived to our times as *Limulus* which is regarded as a "living fossil". The horseshoe crabs together with the extinct eurypterid's of Palaeozoic age are united in one class of animals, the merostomata. The eurypterids body had a long segmented hind part and individual animals could reach a length of 1.5 metres.

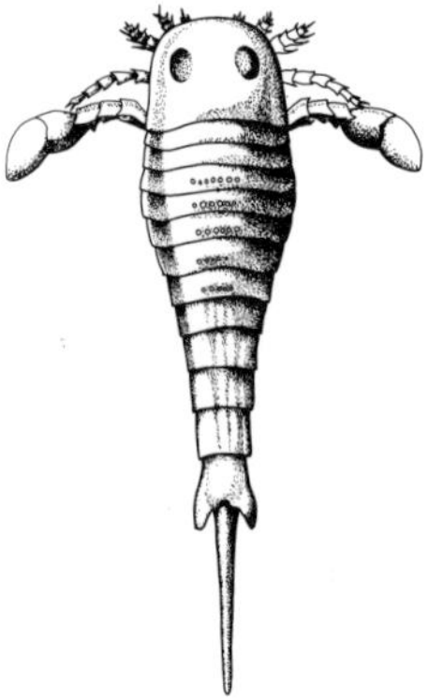

Eurypterid

Crustaceans

114 Short-tailed decapods (crabs) arrived somewhat later than the long-tailed forms. Excellently preserved specimens have been found in Tertiary beds.

115 The tiny ostracodes are genuine crabs which, like mussels, are enclosed in a shell composed of two valves. They have a special importance as a means of identifying the ages of rocks penetrated in petroleum-boreholes. Especially well-preserved decapods are known from the Solnhofen Limestone and the Chalk in Australia. Their modern-day relatives are langoustes and lobsters, on exhibition in Room 204 (2nd floor).

Because of the temporary exhibition of fossils from Messel in the 2nd gallery, the fossil mollusks, the echinoderms and the trace fossils are at present not on display.

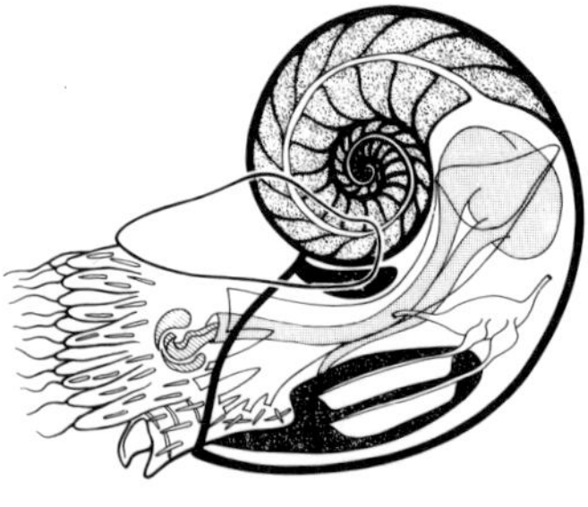

Diagram of a cephalopod

Mollusks

Mollusks have been an important group of marine animals at all times in the past. They have produced a huge variety of form of their shell and have been, at times, massively abundant. The most important classes of mollusks are those represented by the snails, the mussels and the cephalopods. Snails (gastropods) and mussels (bivalves) were already extant at the beginning of Palaeozoic era, but came to their peak development much later. During the Cretaceous some bivalves, the rudists, assumed the cup-like shape of a large coral and built up reefs, others had snail-like, spirally wound shells (Requienia). *Campanile giganteum* stands out among the snails because of its size.

The cephalopods (which include fossil ammonites) are marine animals whose present day representatives include the octopi, the squids, and the pearly boat, the *Nautilus*. There is a rich variety of shell form. Some shells are long, drawn-out, others are like flat, planospirally wound shells of gastropods. A cast of the largest ammonite ever found is on display in front of the museum. Cephalopod shells are always divided internally into separate chambers. The living animal occupied only the last developed chamber, the so-called living chamber. The other chambers were filled with gas, which made it easier for the animal to float in the water. The ammonites are especially important for unravelling the age of Mesozoic rocks. One distinguishes families within this group on the basis of the complex, curved forms of the edges of the walls seperating the chambers. This suture line is, however, only to be seen if the external wall of the shell has been weathered off or removed by special preparation.

An example of using ammonites for defining geological time zonation is shown, based on ceratids in a section of the German Muschelkalk.

In belemnites, the soft parts enclosed the shell and the massive end of the shell (often fancifully identified as "Jove's thunderbolts") can be used to characterize particular time levels in the Jurassic and Createaceous systems. In the Solnhofen Limestone of Late Jurassic time, relatives of

Ammonite

the squids, the so-called teuthoids, have been found with their cuttle bone, ink-sac and body-outline preserved.

Echinoderms

The echinoderms (Echinodermata) live exclusively in the sea. They produce firm calcareous skeletons made up of little plates. One group, the sealilies (Crinoidea), has arms and root-like growths and there is a superficial resemblance to plants (see the two large slabs in the entrance hall). Most of the crinoids are fixed to the bottom by a stem. When the animal dies, most of the individual rounded plates which make up the stem usually fall apart (trochites, in Germany named St. Boniface's pennies) and may be so numerous as to produce what later becomes a crinoidal sedimentary rock.

The sea urchins are another long-established group of echinoderms. They first became abundant and diverse in Mesozoic era. Originally they were radially symmetrical, spheroidal animals (Regulares), which were agile on the sea bottom. The form of their spines is often bizarre, and spines are often found isolated in the rock. In Jurassic time another group (Irregulares) of sea urchins branched off from the regular forms and is now readily distinguishable from the root-stock because its members became bilaterally symmetrical. This group, too, is still represented today. Its living members are the so-called "heart urchins" (*Echinocardium*) which burrow into soft sea bottoms.

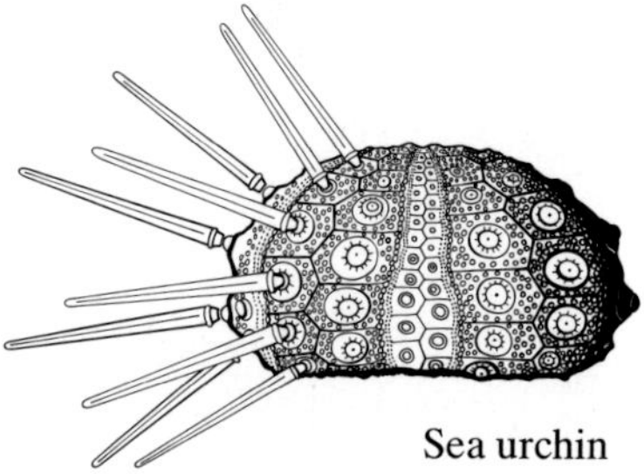

Sea urchin

◀

Large ammonite (*Placenticeras whitfieldi*) with exposed suture;
Below: Two small ammonites of the genus *Cleoniceras* with mother-of-pearl layer preserved

7

Fossil track of a crawling animal

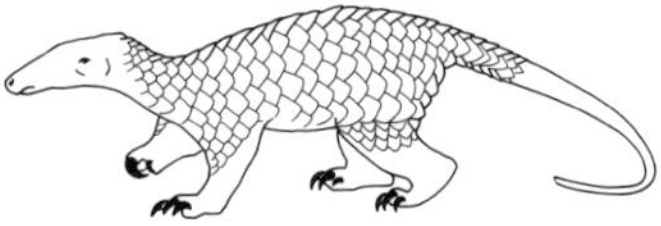

Reconstruction of a fossil pangolin

Primeval artiodactyl

Trace Fossils

Even more remarkable than the fact that details of the bodily form of prehistoric animals have been preserved for us is the fact that delicate traces of their activities during life have survived in the sediment for millions of years. The bedding surfaces of some rocks are entirely covered with such traces, and one can read in them the actions of animals long gone just as a hunter can read tracks left in fresh snow. Thus, traces help the geologist to interpret fossil sea floors. Some of them are readily recognizable as such of worms, snails, bivalves or tracks of crabs, others may be identified as having been left by amphibia, reptiles or mammals. Dinosaur tracks, for example, have yielded interesting details of information on the structure of the foot and on the gait of these large animals. The tracks of an *Iguanodon* from the Cretaceous sandstone can be seen in Room 3.

Messel Fossils

The 2nd gallery of Room 7 shows a temporary exhibition of fossils excavated in the former oil shale pit Messel near Darmstadt. Since 1975 the Research Institute Senckenberg has carried out intensive excavations.

In the sapropelic deposits of a fresh water lake, remains of a flourishing animal and plant life of the Middle Eocene have accumulated. They are preserved in a very rare and unique way. Vertebrates are often preserved as whole skeletons with black shady outlines of skin, pelt and plumage. The fossils come from a time 50 million years ago. At the time the Messel fossils were alive as animals and plants, the Alps did not exist as a mountain belt. Consider the time necessary to build up such a mountain range millimeter by millimeter.

At mid-Eocene time, climate in central Europe was much warmer than today due to different positions of equator and poles. In the Messel area the average annual temperature was 20°C, i. e.

Turtle (*Allaeochelys crassesculpta*)

subtropic to tropic. In the area of Frankfurt at that time palm trees prevailed and on the banks of the Messel Lake alligators idled in the sun and lurked for prey.
The temporary exhibit is restricted to vertebrate fossils excavated between 1975 and 1977. Plant remains, fossil insects and the wealth of verte-

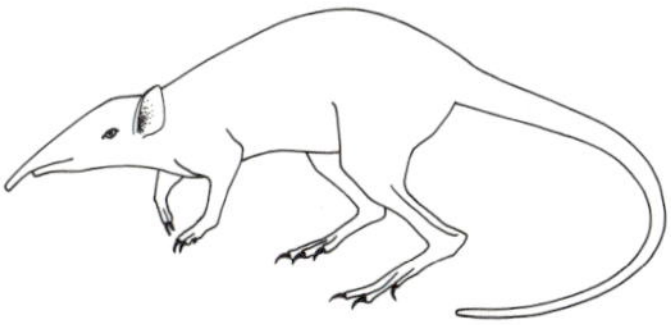

Pseudo elephant shrew

brate fossils accumulated since then are being prepared for a permanent Messel exhibition.
The present exhibition starts with an introduction into history, position and geologic situation of the site. A review of the fish fauna is followed then by small exhibits on amphibians, small reptiles, turtles, snakes, alligators and crocodiles, as well as birds. The mammals, famous for their primeval horses and insectivors also include a variety of bats.
Messel fossils are world-famous not only because of their excellent skeleton and softtissue preservation but also for the preservation of their stomach content. Modern research on these revealed detailed information on mode of life and diet of the Messel animals.

Fossil hedgehog from Messel

◀
Primeval horse (*Propalaeotherium parvulum*)

8

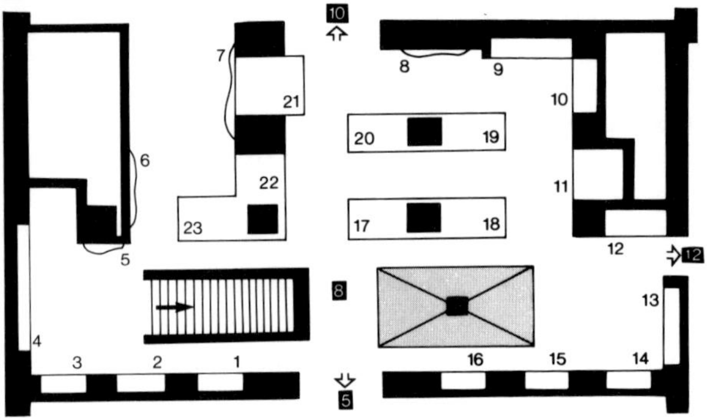

Room 8, Fossil Mammals (under Construction), Vertebrate Fauna from the Mainz Basin (under Construction), Modern Aquatic Mammals, Archaeopteryx (Temporary Exhibition), Reptiles (Provisional)

This room forms the connection between the first and the second halls. At present it still contains several completely different displays. The one half is devoted to the geology and vertebrate fauna of the Mainz Basin, and to Ice Age mammals. The duck-billed dinosaur (*Anatosaurus*) can be viewed from above through a window in the floor. The other half is at present occupied by the *Archaeopteryx*-exhibition and various reptiles. Modern aquatic mammals lead into Room 10, where the whales are displayed. In the course of further expansions to the display collection, this room will be reserved mainly for fossil mammals.

Vertebrate Fauna from the Mainz Basin
So far, only one showcase has been installed for the geological and faunistic evolution of the Mainz Basin. It depicts the Ice Age (Pleistocene)
as exemplified by the Mosbach Sands (approxi- 13
mately 500,000 years old) and their famous mammal fauna. These sands are deposits from the Main and Rhine rivers of the Ice Age, in which the remains of an abundant mammal fauna has been found. Besides badgers, horses, deer, beavers and hares, there still used to be bears, hyenas, sabre-tooth cats, lions, elephants, rhinoceroses, hippopotamuses and monkeys in the Rhine-Main Area at that time.
Further cases are planned to show the environments and fauna in the Upper Pliocene, Upper Miocene, and Lower Miocene epochs.

The cast of a *Deinotherium* skull from the Upper Miocene at Eppelsheim is displayed in Room 10. A considerably older element of the mammal fauna from the Mainz Basin is the hornless
19 rhinoceros (*Dicerorhinus tagicus*) from Budenheim near Mainz. It is a smaller, lightly built precursor of the modern Sumatran rhinoceros (*Dicerorhinus sumatrensis*) which is in extreme danger of extinction. In the Lower Miocene, about 25 million years ago, when the climatic conditions here were warmer than today, this genus also made its home in our latitudes. Even older than the hornless rhinoceros is the
9 fossil sea cow, *Halitherium schinzi*. It comes from the Oligocene marine sands at Alzey, which are estimated to be about 30–35 million years old. At that time, the Mainz Basin was covered by a shallow sea where the sea cow, a slow swimmer, fed on water plants. Next to it is the cast of the much larger skull of Steller's sea cow (*Hydrodamalis gigas*). Although first discovered in the Bering Sea only in 1741, it had already been exterminated by man by 1768. Deposits near Flörsheim also aged between 30 and 35 million years were the finding place of the large skeleton
20 of a marine turtle (*Chelonia gwinneri*). The skeleton of a modern green turtle (*Chelonia mydas*) is displayed without its shell for comparison.

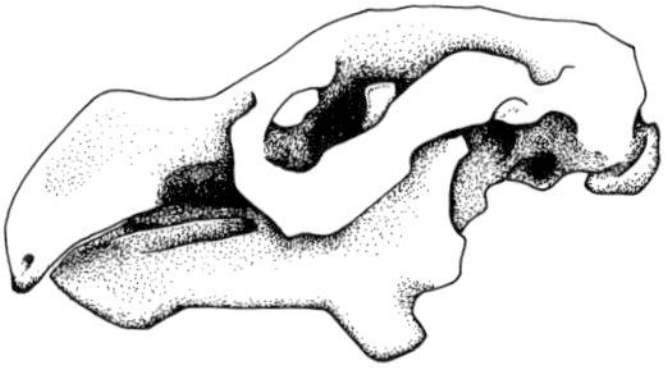
Skull of Steller's sea cow

Ice Age Mammals

11 The cave bear (*Ursus spelaeus*) was widely distributed in Europe south of the northern ice layer in the more recent Ice Age about 180,000–10,000 years ago. Drawings and carvings indicate that it was well-known to primitive man (see Room 13, case 12). Skeletal remains of many thousands of specimens have been dug up in caves where it used to overwinter. The cave bear was larger than its living modern relative, the brown bear, from which it was also distinguished by its highly arched forehead.

Skull of cave bear

The sabre-tooth cat (*Smilodon*) from the asphalt deposits of Rancho La Brea on the outskirts of Los Angeles, California, is around 100,000 years 10
old. Thousands of animals once sank into the deceptive subsoil and so remained preserved. In contrast to other large cats (compare the skull of the modern tiger), sabre-tooth cats as carrion eaters had very long, blade-like canine teeth with which they could slit open the cadavers of dead animals to get at their intestines. Having short hindlegs and a short tail, they were not good jumpers, but rather resembled hyenas in their mode of life, though hyenas, in contrast to the extinct sabre-tooth cats, are more bone eaters.

Sabre-tooth cat (reconstructed)

During the Ice Age, dwarf forms of large mammals developed several times on the islands in the Mediterranean. Examples of this are the dwarf elephants exhibited in Room 10.

Mammals of the Tertiary Period from America 18

Showcase 18 is devoted to the comparison between an early herbivorous and an early carnivorous mammal. Both skeletons were discovered in Lower Eocene sediments in Wyoming (USA) which are dated as 50–55 million years old.

The herbivorous *Phenacodus,* a primitive ungulate, belongs to the primordial group of the Condylarthra. In contrast to modern hoofed animals, *Phenacodus* still had five fully developed toes with small hooves on each leg. With its large canine teeth, stooped posture, toe walk, and long tail it still very much resembled the equally old primitive predators, the Creodonta.

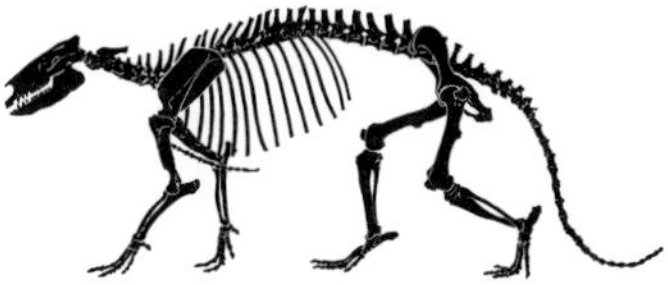

Primeval ungulate *Phenacodus*

The exhibited specimen of this group, *Sinopa rapax,* did not yet possess the carnassial teeth typical of most predators.

The further evolution of the ungulates is represented by *Merychippus,* a three-toed primitive 12
horse from the Middle Miocene of North America. It stands at the beginning of the evolution of the modern horse type about 18 million years ago with the lengthening of the lower sections of the

limbs and the development of high crowns on the molars.

Modern Aquatic Mammals

All mammals adapted to life in water originate from former land mammals. Members from the various groups – such as the reptiles of Mesozoic era (Room 6, text p. 32) – shifted their life cycle more or less completely into water. The animals in question are the whales and dolphins (Room 10, text p. 80), the seals and the sea cows. There are also forms within other groups which have adapted themselves anatomically and ecologically to life on or in water, such as the sea otter from the martens (Room 102, case 6), or
7 the hippopotamus from the artiodactyl order. As a herbivore, the hippopotamus looks for its food on land by night, and passes the day in the water and on the banks. Despite its size and unwieldiness, the hippopotamus is a good diver and swimmer, and can stay underwater for a long time. When it is lying in the water, only its nostrils, eyes and ears show above the surface.

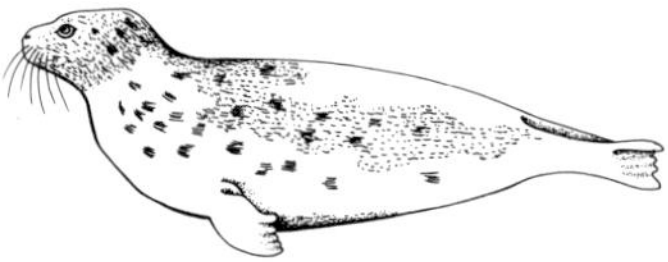

Saddleback

Seals

The seals probably descend from bear-like ancestors. Eared seals and walrusses are numbered among the more primitive seals. They swim by making rowing motions with their arms, which have been modified into flippers, and by undulating their bodies. On land, they can turn their hind flippers forward – which is no longer possible for the more highly specialized seals – and move fairly quickly on all fours.

5 The best known of these is the Californian sea lion, often incorrectly called a seal, an agile animal which dives expertly and swims very nimbly.

6 Among the largest mammals with flippers are the walrusses, inhabiting the Arctic Ocean. They scrape mussels from the seabed with their long tusks, but also eat fish, crustaceans, and many other invertebrates. They are mostly to be found in shallow bays, but also dive to greater depths.

The walrus skeleton exhibited is mounted in a metal frame indicating the outline of the body. The graphic art gives an impression of ice floes with a group of these gregarious animals.

Walrusses

Sea Cows

The sea cows are relatives of the elephants. Their build shows that they – like the whales – are completely adapted to aquatic life: round, streamlined bodies with forelegs modified to resemble fins, no external hindlegs, and horizontal tail fins. At the present time, these herbivorous animals are represented by only two genera living along a few tropical and subtropical coasts and in the inland waters connected with them.

8 The fork-tailed sea cows are represented by a dugong – displayed as a mounted skeleton on a graphic background depicting its habitus – from the Red Sea. Another member of this group, which was exterminated by hunting in historical times, was the more than 7 metres long Steller's sea cow (case 9). From the three species of round-tailed sea cows or manatees, there is the cast of the body of a river manatee from the Amazon to be seen on the display wall, while a lamantin is on show in Room 102, case 19. The lamantin comes from tropical West Africa.

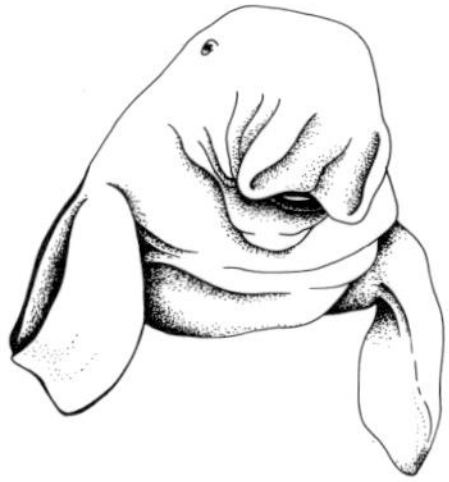

Dugong

Archaeopteryx

This temporary exhibition offers an insight into the life and fossil record of the primitive birds. The history of the evolution of birds and bird flight is depicted by means of casts of all known skeleton finds.

1 The oldest recorded finds, a feather (1860) and the London specimen (1861) form the introduction to the history of fossil finds. Primitive bird feathers were already constructed on the same principles as those of present day birds and probably served as temperature regulators. *Archaeopteryx* is a mosaic form, i. e., it possessed characteristics of both birds and reptiles.

! 8

The topics, biotope and taphocenosis, are combined in the Berlin specimen, which also most
2 clearly exhibits the habitus of primitive birds. The fossils found in Solnhofen come from the most diverse biotopes: open sea, reef areas, mainland coastal strips, and the air. The diversity of these fossils from the lithographic limestone can be admired in Room 9.

3 Plate and counterplate of the Solnhofen and the Teyler specimens are coupled with the problems concerning embedment and preservation. A comparison of the various finds reveals that all of them have the neck bent very much backwards, the tail expanded, and the wings spread out in different ways. Only by rapid embedding in fine calcareous slime could the feathers be preserved as delicate imprints.

Archaeopteryx, Eichstätt specimen

4 The history of the Eichstätt specimen (main slab and counterslab on display) – excavated in 1951, but not recognized as a primitive bird until 1973 – clearly shows that *Archaeopteryx* would have to be classified with the Coelurosauria – so strongly does it resemble them – if its feathers had not been preserved. Coelurosaurs (hollow-boned dinosaurs) were a group of small, predatory dinosaurs widely distributed during the Mesozoic era exemplified by a *Compsognathus* (Room 5) also from the Solnhofen beds. In conclusion various hypotheses on the evolution of bird flight are outlined: the primitive bird forerunner, Proavis, as a flutterer along the ground, as a tree climber with gliding flight, or as a "cursorial bird", which first acquired feathers as protection against heat, and used its wings as enlarged grasping hands to catch its prey.

Reptiles (provisional)
From among the reptiles, the marine turtles, a large land tortoise, and an anaconda in the act of devouring a capybara are being exhibited provisionally in Room 8 at present. All specimens will

be included in the reptile exhibition in Room 104 after further extensions to the museum.

Marine Turtles
The reptiles best adapted to aquatic life are the marine turtles. Their limbs have been modified into paddles; their nostrils, which can be closed, are situated on the upper side of the head. They spend their whole life in the sea, the female alone still going ashore to bury her eggs in the sand. The young turtles hatch after about 100 days and immediately run into the sea where they mature.
The best known marine turtle is the green turtle 22
hunted everywhere for its meat, and threatened by extinction despite its wide distribution in all tropical seas. The horny plates of the dorsal
armor of the hawksbill turtle provide genuine 20
tortoiseshell, though this has fortunately been replaced to a large extent by synthetics today. The heaviest modern turtle is also a sea turtle; at 2 m long and with a weight of 400 kg, the leather-back turtle is a true giant. It differs from other turtles in that its leathery skin does not cover large bone plates but rather a mosaic of small, thick, new bony structures.
The bony armour so characteristic of turtles has been reduced in sea turtles in such a way that it has large open windows at the edges of both the back and the underside. This diminishment, which is essentially a saving of material, was present even in fossil turtles from the Tertiary period. Thanks to the buoyancy of the dense medium water, the supporting skeleton and the armour of these secondary water-dwellers can be built more lightly. The sea turtles of the Atlantic occasionally even stray to German coasts.

Loggerhead

Giant tortoise from Aldabra

Aldabra Giant Tortoise 17
The Giant tortoise from Aldabra and the Seychelles in the Indian Ocean, and the Elephant tortoise from the Galapagos Islands in the Pacific are the largest chelonia living today. Their ances-

tors were once widely distributed over the Earth but they have only been able to survive into the present on these small, out-of-the-way islands. Most tortoise populations are endangered, either because their habitat is being increasingly restricted, or because they are directly hunted for their meat, tortoiseshell, or eggs.

24 **Anaconda**

Since boas are non-venomous, they are dangerous to man at the most due to their extraordinary strength since they kill their prey – vertebrates up to the size of gazelles – by encoiling and suffocating it to death. The anaconda is the largest aquatic snake of South America and shows how widely the mouth can be stetched and what comparatively huge prey can be swallowed whole. The jawbones and palatal bones can move far apart from each other as they are connected by stretchable bands only. During the swallowing process, the larynx, which is stiffened with rings of cartilage, protrudes from the mouth so that the snake can continue to breathe. The prey is a capybara (Room 102, case 21), the largest South American rodent.

Room 9, Trachodon Cellar

The building of extensions in 1963/64 brought an opportunity to add this cellar to the exhibition space available in the museum. It is therefore now possible to view "duckbill", one of the most valuable original specimens in the museum, not only from the sides, but also from room 8 above.

Anatosaurus (Duckbilled lizard, trachodon) **12**
The *Anatosaurus* is one of the ornithischian dinosaurs. Of the four nearly complete skeletons that have been found (in North America) this is the only one on display in Europe. It comes from sandstones in the Laramie Beds of Wyoming which were formed as river deposits toward the end of Cretaceous time. The trachodon shows some special features which allow us to draw some conclusions on the mode of life and manner of death of this powerful reptile. It is astonishing to find that in addition to the animal's skeleton, impressions of its skin and the horny beak have been preserved. This is due to the fact that the cadaver dried out and was therefore to some extent "mummified" before burial. There is an indication of this in, among other things, the twisted neck – in mummification the neck becomes bent and twisted as the neck musculature dries out and tightens (shrinks). The impression of the skin is clearly seen on the forefeet and on the area of the animal which was laid into contact with the sand. The skin of the forefeet looks like a web and the broad, horny beak has the appearance of the sieving beak of a water bird. It was therefore suggested that the trachodon lived in shallow waters and fed on water plants. Its stomach contents, which were encountered during preparation as a brownish mass lying in the body-cavity of the animal, are not entirely in accord with that suggestion. They include remains of branches of conifers. Together with the strong hind legs this may, however, indicate life

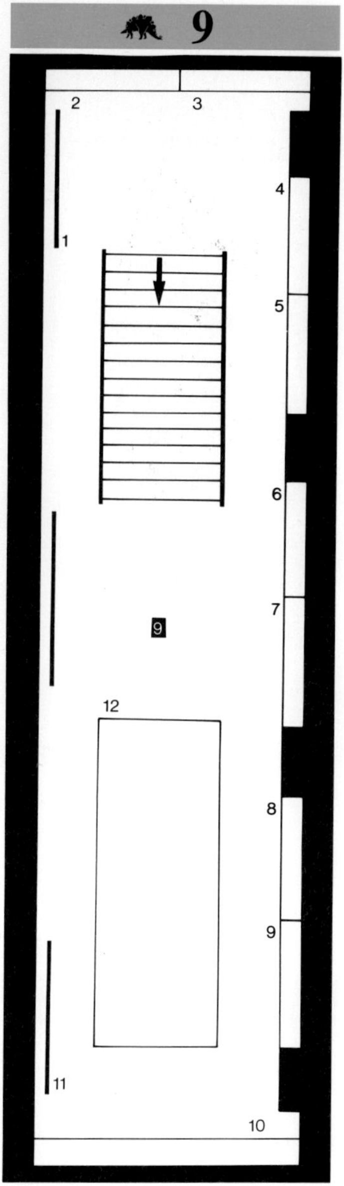

Duckbilled lizard (*Anatosaurus annectens*)

Flying lizard (*Pterodactylus antiquus*) ▶

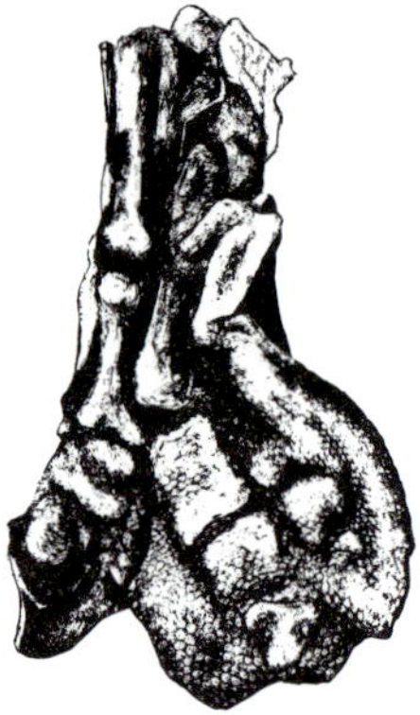

Forefoot with impressions of the skin

on land. The dentition of the trachodon is made up of around 2,300 small teeth.

Lithographic Limestone of Solnhofen

With the exception of the trachodon, all of the fossils in this room come from the Solnhofen Limestone. These limestones first accumulated during latest Jurassic period in shallow marine pools or lagoons and were at that time composed of the finest calcareous mud. Mined as "lithographic shales", near Solnhofen and Eichstätt in particular, the Solnhofen Limestone has become famous throughout the world for its magnificent fossils. They are not merely especially fine and

well-preserved, some of them deserve to be regarded as quite extraordinary palaeontological documents. The assembly of animals in the Solnhofen Limestone represents a number of different environments: there are flying animals and land animals here as well as inhabitants of the lagoons and the open sea, all brought together in a community of the dead (taphocoenosis).

2 **Flying Lizards** (Pterosauria)

Flying forms had already existed at the beginning of Jurassic time among the reptiles: the flying lizards. Their organs of flight were large wing-webs stretched out between the body and the much elongated fourth finger. The first three fingers in the hand were small and bore claws which helped the animal to cling firmly to a perch. The flying reptiles were certainly capable performers in flight. They had a brain in some respects similar to that of the modern bird, had powerful flying muscles and had hair to help regulate the body temperature. The earliest of them in line of evolution, the long-tailed flying reptiles of Jurassic time, had a small rhombic rudder at the end of their tail. Among the long-tailed flying reptiles was *Rhamphorhynchus* ("Beak-nose"), one of the finest of the original specimens of which is on display here, together with a cast of a wing with the web preserved. The short-tailed flying lizards made their appearance in Late Jurassic time and evolved to produce giant forms during the Cretaceous, one of which was *Pteranodon,* which had a wing span of about 8 metres. The short-tailed flying lizard on display, *Pterodactylus* ("Wing finger"), comes from the Solnhofen Limestone and was about pigeon-size. Most of the species of flying reptiles of which we have found remnants probably lived

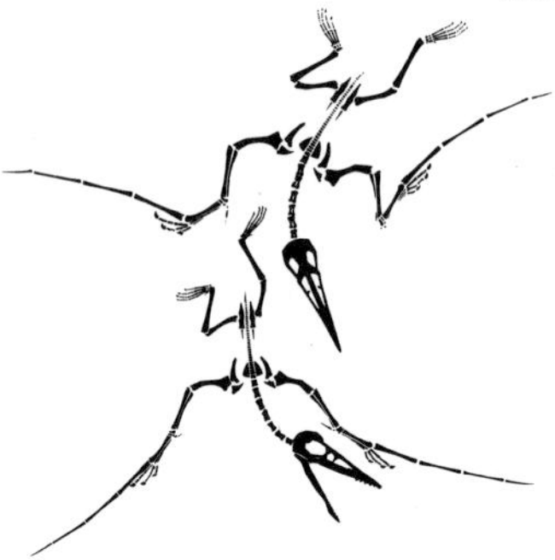

Movement study of a pterosaur

Archaeopteryx, cast of the Berlin specimen ▶

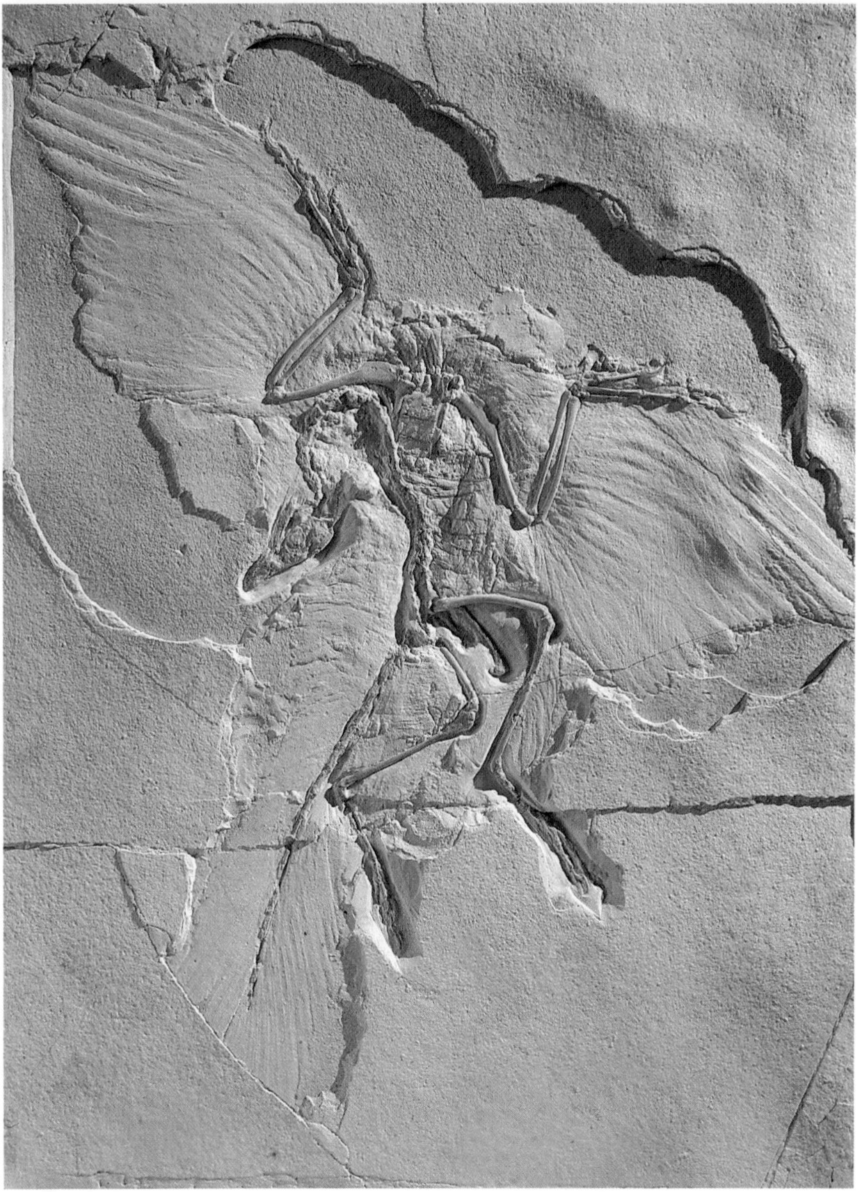

rather in the style of present day coastal and sea-birds. Fish, molluscs and small crabs made up the major part of their food, and may have been caught either by a flying attack or else by gliding over the sea surface.

3 **Archaeopteryx**

The Solnhofen Limestone fossil with the greatest evolutionary significance is the primitive bird *Archaeopteryx lithographica.* Since 1857, a total of 5 skeletons and one feather of this unique animal have been found. We find a combination of reptilian and avian features in its anatomy. The displays here are casts of the two most famous original specimens, the "London" and the "Berlin". They clearly reveal that this creature had on the one hand a skeleton which is in many respects similar to that of the small predatory dinosaurs but that on the other, it had feathers like a bird. This "feathered reptile" is therefore the incorporation of a link between two classes of vertebrates and no other comparably clear example is known among fossils. One notes in the skeleton the reptilian skull, with toothed jaws, the simple ribs and the belly ribs, the vertebral column which shows no fusion in the pelvic region and which continues into a long tail and one notes too the claws on the digits. These reptilian characteristics contrast with the avian features present, the most obvious of which is the feathering. (Temporary exhibition on *Archaeopteryx,* Room 8; model of *Archaeopteryx,* Room 104.)

4 **Reptiles**

We know other reptiles from the Solnhofen Limestone in addition to the flying forms. Among them is *Homoeosaurus,* a primitive land-lizard. It belongs to the beak-headed reptiles (Rhynchocephalia) and is a close relative of the tuatara (*Sphenodon*) from New Zealand (Room 104). It is possible that *Pleurosaurus* with its elongate, almost snake-like body, is also a rhynchocephalian. It has short but powerful legs, a decidedly short neck and a very long tail. One might suggest that these represent adaptation to life at the sea shore. Crocodiles are rare in the Solnhofen Limestone. A skull of a marine crocodile (*Geosaurus*) is on display. This form was closely related to the *Metriorhynchus* shown in Room 6. Like it, this form had paddle-like limbs and a powerful tail including a tail fin with two flukes. Turtles were already extant during the Triassic but it was only in the Late Jurassic that certain groups first became adapted to life in the sea. *Plesiochelys* is a representative of the oldest group of turtles (Amphichelydia), which is now extinct. *Eurysternum* on the other hand was an early cryptodiran turtle. These have been in evidence since late Jurassic time and are still the most successful group of turtles today, with forms inhabiting the land, or swamps, or the seas.

Fish

The commonest fossil vertebrates in the Solnhofen Limestone are fish. Sharks and rays, i.e.
representatives of the cartilaginous fishes, were 5
not at all uncommon in the seas of Jurassic time. The angel fish (*Squatina*) on display is a shark, but it has the form of a ray. One can still see around the intact skeleton the outline of the skin, the wing-like pectoral fins, the small dorsal fins displaced during burial and the long drawn-out tail. There are similar angel fish today, see Room 208.

Ophiopsis was a simple bony ganoid with a complete ring of bone around each eye, a huge gaping mouth and a sail-like anterior elaboration of the dorsal fin. *Urocles,* like the modern *Amia,* had small cycloid scales and was a form intermediate between the bony ganoids and the true bony
fishes. The most abundant among them is *Lep-* 6
tolepis, a genus of fish whose size is in the sprat to herring range and which lived in swarms. These
were true bony fishes (Teleostei), related to the 7

Fossil angel fish

modern herring. *Thrissops,* too, belonged to the herring-like fish, a group which comprised the root-stock of the true bony fishes and which had already appeared in Triassic time, although it was in the Jurassic that the group began to flourish. *Thrissops* had a deeply divided tail fin and its jaws bore powerful teeth. It was a fast-swimming predator. The other fish on display all belong to the bony ganoids (Holostei). *Gyrodus* had grasping teeth and battery of bullet-headed teeth in its mouth. The ridgelike thickenings of the anterior margins of the ganoid scales have the appearance of a lattice work. The body form and dentition suggest that this fish lived among coral reefs and fed on hard-shelled invertebrates.

Orthocormus, Caturus, Aspidorhynchus and *Be-* 8
lonostomus were large predatory fish and good 9
swimmers. The last two had strongly projecting snouts.

Lepidotes was armoured with large, shiny rhom- 10
boid scales. Projecting grasping teeth and a battery of bullet-headed teeth served to break down hard-shelled food material such as bivalves. Similar adaptations are seen in the placodonts (Room 6) and the sea otter. The two *Lepidotes* specimens on display are the only ones of their size. One was found already broken down into fragments, with the rock in which it had lain in a like state. It proved possible to cement all of the pieces together again and then to chisel the fish out of this reconstituted bed of rock.

Further fossil fish are to be seen in Room 6; a survey of the evolution of the various groups of fishes is given in Room 208, 2nd floor.

◄

Giant *Lepidotes maximus* with bulled-headed teeth

10

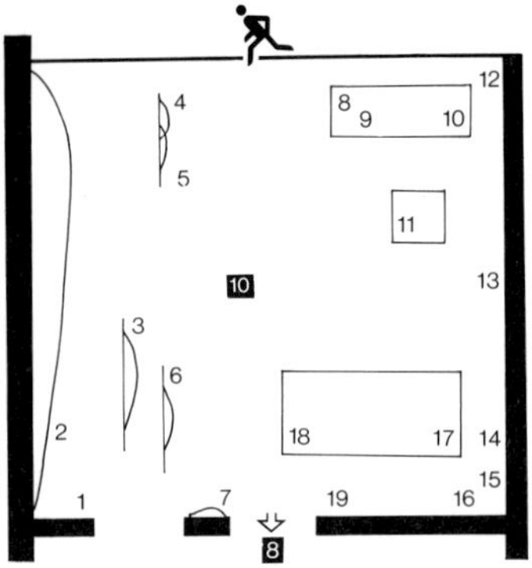

Room 10, Second Hall, Whales and Elephants

The second hall is devoted to mammals living at present and their fossil relatives.

Whales

Whales are marine mammals descended from primitive ungulates (hoofed animals), but completely adapted to living in water. Like all mammals, they are warm-blooded, breath air from the atmosphere into their lungs, and the mother animals suckle their young which are born alive. A fundamental distinction is drawn between baleen whales and toothed whales.

Baleen Whales

The largest animals of all times belong to the baleen whales. They filter relatively small animals out of the water as their food using a straining apparatus, the whalebone, a flexible horny substance attached to their upper jaw. The second largest species of the group, the finback whale, is the only baleen whale exhibited. It occurs in all oceans of the world, but prefers the Arctic and Antarctic seas.

2 The skeleton of a 22 metres long finback whale (*Balaenoptera*) is mounted on the south wall against a cement cast which visualizes the outline of the massive body.

The baleen whales are in extreme danger. Modern whaling ships kill from 60,000 to 70,000 whales annually, which further decimates the stock from year to year. Many a species, which formerly populated the oceans in large numbers, has become rare today. The complete extinction of some species, of which there are only a few specimens still alive, can probably no longer be averted in spite of the introduction of closed seasons and other protective international agreements. The extermination of the blue whale, finback whale, and sei whale seems to be inevit-

Pilot whale (*Globicephala melaena*) and on the wall (background) finback whale (*Balenoptera physalus*)

able. In earlier times, baleen whales were harpooned by hand and later with compressed air harpoons, today explosive and electric harpoons are used. The animals are located with helicopters and ultrasound devices, captured from aboard fast fishing boats, and processed immediately in floating blubber factories. Within half an hour, the blubber and oil is removed from a whale weighing 50–90 tons, and everything is packed in cans and barrels. The mother ship of a whaling fleet processes 50 whales and more a day.

Toothed Whales

The toothed whales have teeth instead of whalebones. One of these is the 6–9 metres long killer
3 whale which is a predator on large animals. It inhabits all seas, and is occasionally even observed in the North Sea. Often hunting in packs of 20–30 animals, it lives mainly on fish, seals, porpoises, and dolphins, but can also tear large baleen whales apart.

6 Another toothed whale is the narwhal, the "unicorn of the sea". Its teeth are small and mostly still covered by the gums. The male has a large tooth which protrudes horizontally from the upper jaw, and can be up to 2.5 metres long. Strangely, it is almost always the left tooth which grows to such an enormous length while the right one remains undeveloped as a rule. Narwhals with two tusks are extremely rare. However, only the males have this tooth development; the females have no tusks. This is an example of sex dimorphism. The function of the tusk is unclear; it probably plays a part in the whale's social behavior. Narwhals seldom grow to more than 6 metres in length. They live in the seas in the far north, going the farthest north of all whales. Only rarely does such a "unicorn" enter the North Sea. Their food consists mainly of animals living on the seafloor such as rays, flounders, crustaceans, and even sea cucumbers, but they do eat free swimming marine animals as well. The pilot
5 whale also grows to a length of almost 6 metres. Its head is thickened to a spherical shape. It lives in large herds in the world's oceans and feeds on cuttlefish and fish.

Killer whale

Narwhal

The bottle-nosed whale is a toothed whale, too. Its teeth, however, are undeveloped and its snout is shaped like a birdbill. 4 The few teeth it has scarcely penetrate the gums, and the old animals lose them completely. The head of the males of this species is very much distended. The upper jawbones form massive longitudinal ridges containing oil-filled cavities. The preferred food of the bottle-nosed whale consists of cuttlefish and herrings. Bottle-nosed whales live in the northern Atlantic including the North Sea and reach a length of up to 10 metres.

On a concrete pillar on the east wall, you can see sculptured replicas of a bottle-nosed dolphin. 7 They come from an injured animal caught in the Blaue Balje (Jade) near Wilhemshaven by the research ship of the Senckenberg Research Institute. These bottle-nosed dolphins are the smallest dolphins, and used to be the most frequent ones in the North Sea; the present stock has been very much reduced by water pollution. They feed on young herrings and sprats, but also eat cuttlefish and small crustaceans. They live both solitarily and in small bands and are often found in the direct vicinity of beaches.

Only four species of toothed whales have adapted to live in freshwater. One of these, the Amazonia dolphin, is displayed in case 19, Room 102.

Elephant Group

The large land mammals in the elephant group form a complement to the large marine mammals. Representatives of all large groups of the proboscideans (animals with trunks) are on display. To get a complete overview, you should not miss visiting the proboscideans' nearest relatives as you continue your way through the museum. They are the dugongs or sea cows (Room 8), and – surprizing as it may sound – the hyraxes (Room 102, case 15), which outwardly resemble the marmot rather than the elephant. Nevertheless, fossil finds, shared anatomical features, and

Hyrax

Elephant group. In front American mastodon and the Mühldorf mastodon; in the back dwarf elephants and skeleton of an female Indian elephant

serological studies have shown that these externally so different animals descend from a common precursor, and have evolved from the prehistoric ungulates (Room 8, case 18). So, proboscideans and ungulates together form the superorder Paenungulata.
The feature all proboscideans share is the trunk formed from the nose and the upper lip. In earlier forms it was short and functioned as a counterpart to the incisors on the lower jaw in

gathering food. Then, in the course of evolution, it rapidly became longer and developed into the sole organ for grasping food.

A further characteristic feature of the proboscideans is the reshaping of the incisors into tusks – two each in the upper and lower jaws. In the course of evolution, the lower jaw tusks became smaller, and have disappeared in the genera that are younger in the earth's history.

Phylogenetic Tree of the Proboscideans 12

The various lines of development of the proboscideans are depicted in the most modern scientific way on a 24 metres long graphic display wall. Reproductions truly scaled down to 1 : 5 enable you to make direct comparisons between the forms and sizes of these groups.

The phylogenetic tree shows three separate lines of development traceable back to three small forms in the Oligocene epoch which were found in the Fayum Oasis (Egypt). Due to the lack of tusks in the upper jaw, and the reshaping of the lower jaw tusks into huge fangs with downward curvature, the deinotherium is not closely related to the other two lines of development of the mastodons (including the elephant).

The mastodons are descendents of the *Palaeomastodon* and *Phiomia* which were characterized by a smaller body size, a low, long skull, and two short tusks in each of the upper and lower jaws. In both of these two genera, there are already indications of the distinguishing feature of the two groups, the zygodont (i. e., with the cusps of the tooth united in pairs) molars in the first, and the bunodont (i. e., with low conical cusps) molars in the second. The line of zygodont mastodons evolved from the *Palaeomastodon;* the lower jaw tusks became smaller, and the skull higher and shorter which led to forms like the American mastodon. The much more numerous forms of the bunodont mastodon evolved from *Phiomia.* Numerous groups developed which are disting-

Palaeomastodon

Phiomia

uished from each other mainly by the skull shape, the number and shape of tusks, and the structure of the molars. It was mainly the development of more and more efficient types of molars, which made it possible to exploit hard plants as diet, that created the prerequisite for this animal group to colonize all climatic regions of the Earth. Finally, during the Pliocene, the mammoths and our modern elephants evolved from them by retrogression of the lower jaw tusks and the development of lamellar teeth with high crowns.

The two species still living, the African and the Indian elephant, are the last survivors of a once large animal group with many forms.

The colored columns on the phylogenetic tree depict the lines for which there is evidence in fossil finds. The white connecting lines between the phylogenetic lines indicate the probable relationships between the various proboscidean groups, i. e., they are interpretations of the scientific findings corresponding to the present state of knowledge. So our phylogenetic tree – like every other one, by the way – combines objective evidence and subjective interpretation.

15 **Tooth Evolution**

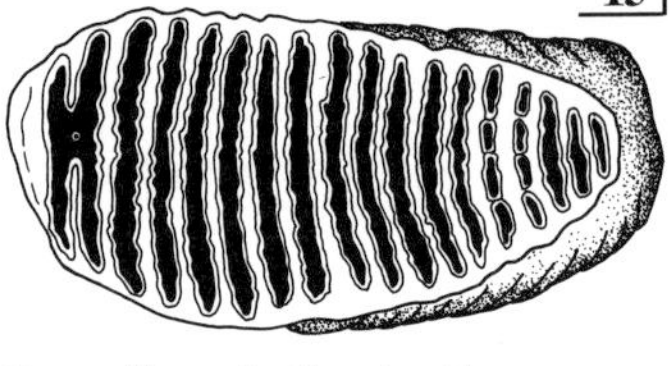

Lamellar grinding tooth

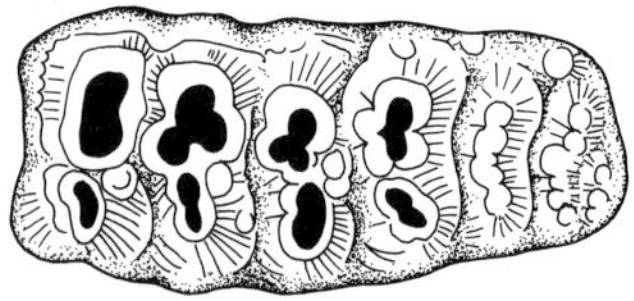

Zygodont crushing tooth

Next to the phylogenetic tree, there is a table depicting the development of the molars of the proboscideans, in which a simplified version of the phylogenetic tree can be discerned. The depiction shows that the phylogenetic tree of the proboscideans is essentially based on a detailed study of their teeth. Two fundamentally different types of teeth occur: the grinding teeth of the bunodont proboscideans colored in reddish-brown, and the crushing teeth of the zygodont proboscideans colored in green. The nuances in the basic color reflect the number of transverse ridges on the molars. This provides the ultimate key to understanding the phylogenetic tree where the colors of the phylogenetic lines correspond to those on the table depicting the teeth, and so indicate the structure of the molars of these animals.

16 **Distribution of the Proboscideans**

The large world map on the east wall of the hall shows how the proboscideans spread out in two great migratory waves. Only the most important groups are depicted to avoid making the picture confusing. The first dispersal wave was that of the mastodons at the beginning of the Neogene (about 25 million years ago) starting from their original home in Africa, covering the whole of Eurasia and reaching America. The green arrows symbolize the migratory routes of the zygodont mastodons, and the dark reddish-brown ones the routes of the bunodont mastodons. The genuine elephants spread out in a second wave, which largely followed the old routes taken by their ancestors, the bunodont mastodons from Africa. The light reddish-brown arrows indicating this migration are distinguished from the other arrows on the world map by being raised so as to emphasize the lapse of time separating this later migratory wave in the Pleistocene (about 2 million years ago) from the first one.

African and Indian Elephants

9 The African elephant is the largest land mammal and attains a body weight of 5–7.5 tons when fully grown. Although forced out of large parts of its former habitat area, its further existence is assured. Elephants' tusks correspond to the incisors of other animals. They consist of dentine only, have no enamel, and reach a length of 2.5 metres and a weight of over 50 kilograms in adult African elephant bulls.

10 They are rarely longer than 1.5 metres and heavier than 20 kilograms in Indian elephants; the cows of this species have only very small tusks if any at all. Elephants' molars have a biting surface like a washboard with many ridges of

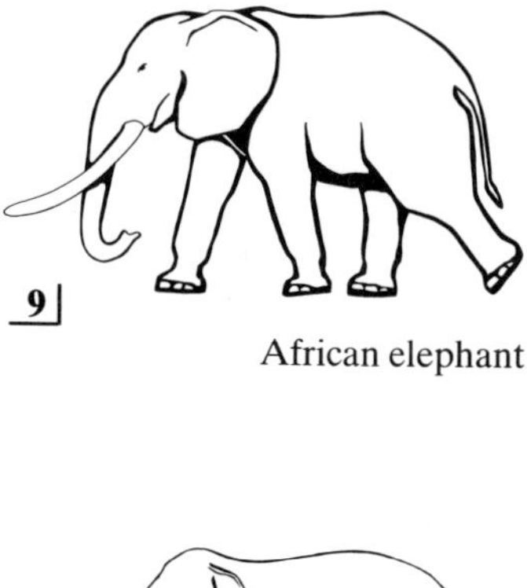

African elephant

Indian elephant

enamel. In the course of an elephant's life, one molar after another appears in each jawbone, and is gradually worn down. The last one is used up at the age of about 65 years, which sets a limit to the elephant's life span.

11 **Dwarf Elephants**

During the Ice Age, dwarfed forms of large mammals developed repeatedly on islands in the Mediterranean. The restricted living space and a deterioration in the food situation probably played causal roles in this development. At the same time, some small mammals grew to giant size which might have been due to the lack of predators and to their occupation of niches which had become vacant. The fully grown pair and the skull of a young animal come from deposits in a cave near Spinagallo (Sicily) which are about 70,000 years old.

Mammoths

13 The steppe mammoth lived during the younger Ice Age and their presence in Central Europe can be demonstrated from the Elster (Mindel) glaciation onwards. They are direct ancestors of the woolly mammoth which developed from them at the beginning of the last glaciation period.

The steppe mammoth is a transitional form linking the older Southern elephant and the woolly mammoth. Early forms display resemblances to the Southern elephant, while later finds have similarities with the woolly mammoth. In this evolutionary trend, the number of lamellae and the height of the crown of the molars increase, which can be interpreted as a continuous specialization in ever harder plant food. The outline of the skull became clearly angular, which facilitates the distinction of the steppe mammoth from the woolly mammoth that has a rounded skull dome.

The genuine mammoth of the last Ice Age, the woolly mammoth, was widely distributed in Germany also, as is proved by numerous finds of

Steppe mammoth

molars as well as of other skeleton parts. The woolly mammoth was a contemporary of the Neandertal and Cro-Magnon man as cave paintings, drawings scratched into rocks, and ivory sculptures show (see Room 13, case 12). With a shoulder height of up to 3 metres, the woolly mammoth was somewhat smaller than our modern African elephant. The woolly mammoths became extinct about 10,000 years ago in their last refuge, Siberia, where many carcasses have been found in the permafrost. Studies of their stomach contents have revealed that their food was mainly tundra vegetation, but also branches of pinetrees, alders, willows and birchtrees.

19 Specimens of woolly mammoth remains, which were preserved in the Siberian permafrost, are displayed in a showcase. These are bristles, stomach contents, blood, and pieces of tissue from the Beresovka mammoth, so called after the Beresovka River in Siberia where they were found.

Mastodons

One typical representative each from the bunodont and the zygodont species is exhibited. The American mastodon is one of the zygodont elephants, which were distinguished by having simple crushing teeth. When chewing, the transverse ridges in the lower jaw teeth fitted into the troughs in the upper jaw teeth and vice-versa so that the food uptake was crushed to pieces between them.

18 The skeleton of the American mastodon on display is younger than 10,000 years. This means that the animal lived at a time when the Indians had already settled in America. The original skeleton is one of the most valuable possessions of our museum. Only a few pieces have been added to it, and so it is one of the most complete skeletons found so far.

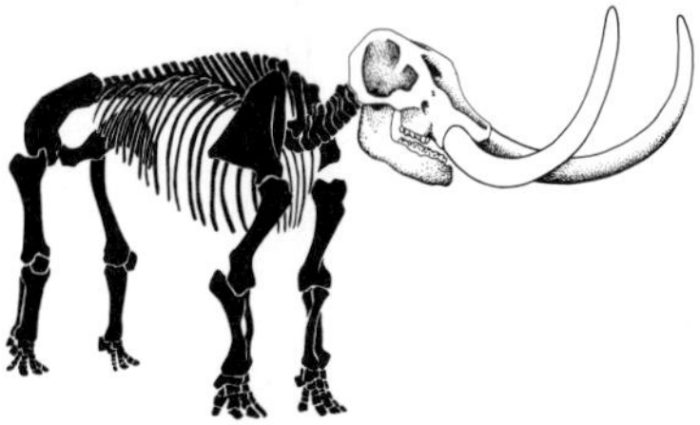

American mastodon

17 The Mühldorf mastodon (*Gomphotherium*) is one of the bunodont mastodons, and is rep-

resented here by the oldest and largest exhibited skeleton. A true primeval giant! The original skeleton, which is kept in the Bavarian State Collection in Munich, was discovered in Miocene river sands on the banks of the Inn south of Mühldorf in 1971, and is about 10 million years old. Because of its size and completeness, it is a unique find in Europe. Especially conspicuous is the long, lowset skull with 4 tusks in all, 2 in the upper and 2 in the lower jaw. This, and the simple structure of its molars, makes the *Gomphotherium* recognizable as a very primitive, still hardly specialized member of the proboscideans.

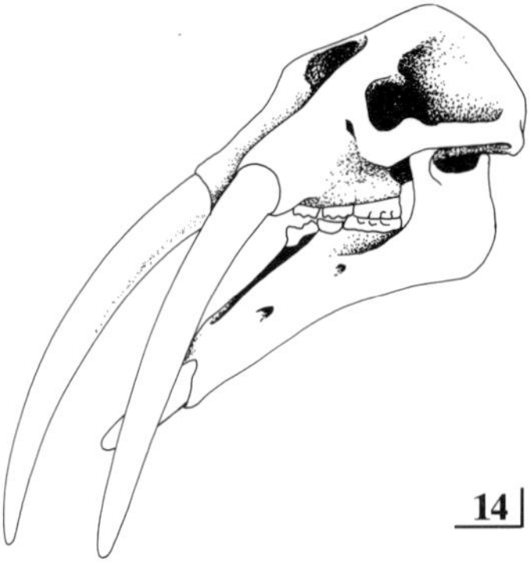

Skull of Mühldorf mastodon

14 **Deinotherium** ("Hoe-tusker")

The *Deinotherium* has simply structured zygodont molars. Its lower jaw tusks were reshaped to mighty downwardly curved fangs while the upper jaw tusks were missing. The low skull has a strikingly large nasal opening which is evidence of a pronounced trunk. The originals of the skull cast exhibited were found in the Upper Miocene sands at Eppelsheim and are about 11 million years old.

A separate brochure in German with detailed information on the presentation of the fossil proboscideans and the individual exhibits has appeared as the loose leaf folder No. 2: "Unsere Elefantengruppe im 2. Lichthof" in German, which can be obtained at the admission desk or ordered from the office of the Senckenberg Nature Research Society.

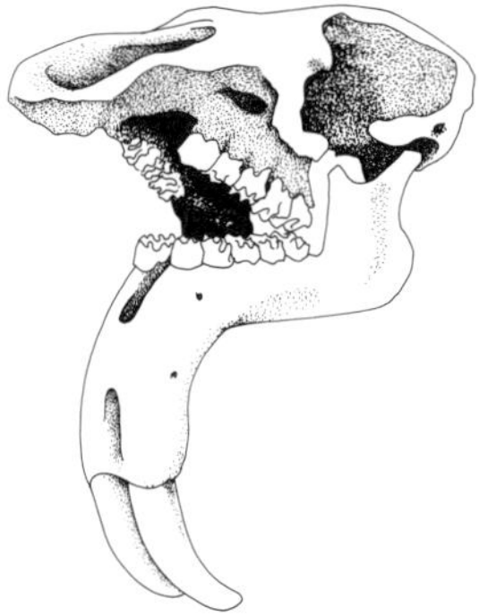

Skull of "hoe-tusker"

Room 12–13, Origin of Man.

Room 12, Biological Affinity of Man and Fossils Wayside to Man.

What is man?

The biological point of view is that man is a mammal. Man is, together with lemurs, monkeys and apes a member of the Primate order. The great Swedish naturalist and systematist Linnaeus had already recognized this in the middle of the eighteenth century. He passed to man the problem of diagnosing his own condition: "Nosce te ipsum – know thyself." Linnaeus accepted the biblical story of the Creation literally, and so it never occurred to him that within his systematic subdividisions there might lie hidden more than one generation of form, their interrelationships to be detected in similarities of structure. Ever since 1859, when Darwin opened up new realms of thought with his work "On the Origin of Species", we have understood that genuine correspondences in the anatomical and chemical make-up of organisms are more or less evidence of common ancestry. Thus we recognise that of 1065 anatomical characteristics we have 113 in common with the Old World monkeys, 354 in common with the orang-utan, 385 with the gorilla, and as many as 396 with the chimpanzee. It is, for example, only with the apes that we can compare the pattern of the grinding surface of the check teeth in the lower jaw. Five cusps – three outer and two inner – define a pattern of furrows in the form of a Y. We presume that we, and the apes, have inherited this pattern from a common ancestor. This consideration brings our attention to a group of fossil apes, the dryopithecines, which show perfect development of this "*Dryopithecus*-pattern" named after them. The dryopithecines lived about 20–7 million years ago in Africa, Asia and Europe. The common origin of

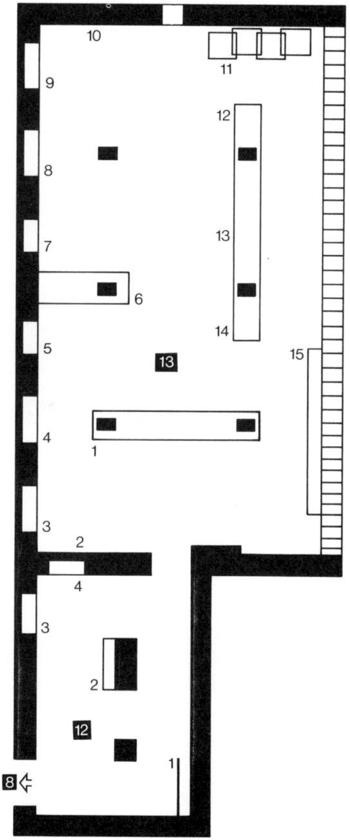

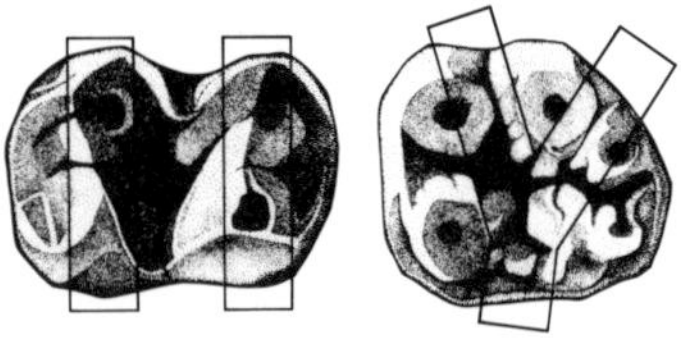

Pattern of grinding surface of lower molars: old world monkeys (left) and apes and man (right)

man and the living apes should be sought within this quite variable group. Taken altogether, the present primates, including man, can be seen as a ladder whose steps represent progressively higher levels of organization, incorporating stages of evolution some of which we have passed through in the course of our own development. Thus the brain becomes progressively larger. The projecting snout seen in the primitive forms is steadily reduced, and in context with that the sense of smell is reduced while sence of sight becomes progressively more important. The eyes still directed to the sides in the tree shrews (*Tupaia*), are shifting to the front so that the two fields of vision overlap and a sense of depth is acquired, well suited to the needs of life among trees.
The suggestion that man is something more than a highly specialized "naked ape" is symbolized by Eve with the apple from the "Tree of Knowledge". It is man's capacity for thought and the special level of operation of his brain that single him out from the animals.

3 **How, in fact, did man evolve?**

We know that in the course of Tertiary time, from about 50 million years ago onward, the world climate has become cooler. A consequent change during the last 20 million years or so has been the spread of open savannah and steppe regions at the expense of jungle regions dominant before. New habitats came into being. Animal life reacted and new adaptations emerged. At the end of the Miocene, approximately 8–5 million years ago, there was a divide among man-like (Hominoidea) animals, separating those which clung to the traditional mode of life in the jungle from those whose evolution can be seen as a response to the new challenges and opportunities offered by the savannahs and steppes. Our ancestors belonged to this latter group. In this almost treeless environment, movement from place to place was no longer a matter for the arms but was done upright on two legs. This group therefore, in contrast to the apes, has lengthened legs. Thus a far-reaching stride was obtained. In keeping with the upright gait, the elongate, shallow pelvis was transformed into a broad basin, which on the one hand supported the internal organs and on the other provided a changed arrangement of muscle attachments better to accomodate the new play of muscles. At the same time the gripping hand, an adaptation to life among trees, was freed from its involvement in the function of movement and now became a specialized organ capable of employing tools and, later, capable of producing tools. With deliberate production of tools, the step from animal to man was made. The decisive development was, however, less the capabilities of the hand – apes, too can seize on objects and use them as tools – than the capabilities of the brain: obvioulsy, the various quick responses of the hand must have stimulated the development of the brain and vice versa.

The australopithecines, found in southern and eastern Africa, make it clear that the sequence of events in human evolution brought first the development of an upright posture then, only later, development of the skull and the brain. These australopithecines were primeval humans, already provided with an indisputably humanoid pelvis (upright gait), but with a brain whose volume of approximately 500 ccm is comparable with that of an ape (a modern human brain has a volume between 1200 and 1800 ccm). The australopithecines first appeared before the beginning of the Pleistocene (Ice Age), about 4 million years ago during the Pliocene. Evolution of man up to this point had been mainly a matter of achieving an upright gait, with attendant development of the pelvis and the lower extremities. At the same time man-like dentition evolved: the canine teeth grew shorter and the diastema – gap between canine teeth and incisors in the upper jaw – became narrow; the two

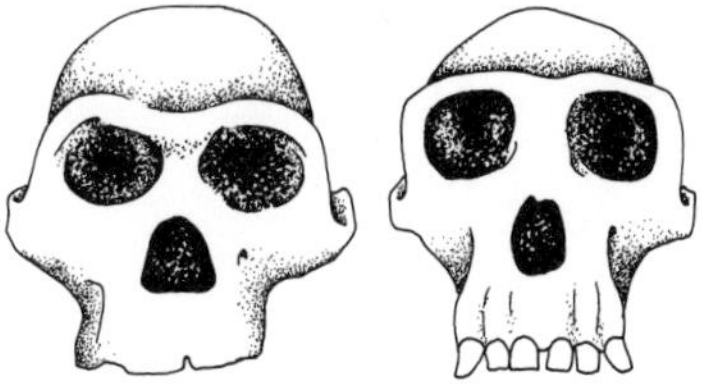

Skull of *Australopithecus africanus* (left) and chimpanzee (right)

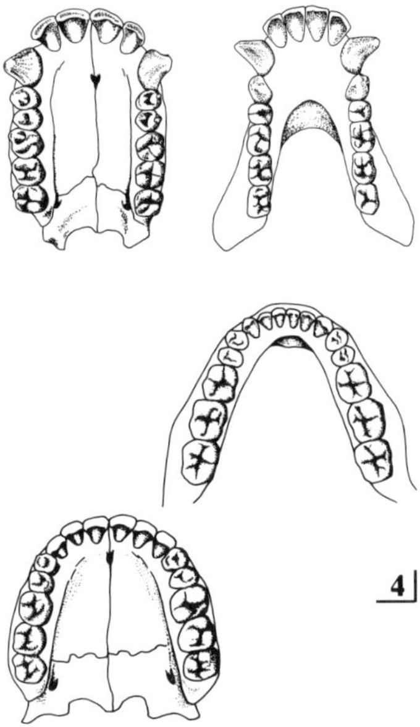

Upper and lower jaws of chimpanzee (above) and modern man (below)

straight rows of chewing teeth changed to a rounded arc.

Human evolution during Pleistocene time and up to the present has been above all the process of developing the skull and the brain to a point at which the brain-volume has become three times that of the australopithecines.

The evolution of man has proceeded in the setting of background developments in earth history. There is evidence to suggest that the whole age of the earth is of the order of 5,000 million years. The oldest fossils known – resembling modern bluegreen algae – have been found in southern Africa in beds of rock approximately 3,200 million years old. The separation of man from apes was, by contrast, achieved as recently as 8–5 million years ago. Man, seen in the context of the whole history of the earth, is therefore a very recent arrival on the scene.

4 **How does one obtain such information on age?**

There are two basically different means of determining age: the relative and the absolute. Relative age-determination only tells us whether one find of fossils, or a rock, is older or younger by comparison with another, or is of about the same age. Such a judgement is based on the stage of evolution reached by the fossil plants or animals or on the order of succession of rocks (older rocks below, younger above or else younger igneous rocks intrusive into older). Absolute datings, on the other hand, provide statements of the age in numbers of years. This can be done using the regular rate of radioactive decay of certain elements, which proceeds naturally in some minerals within rocks (e. g. potassium/argon in volcanic rocks), or by measuring ratio of carbon isotopes, which can sometimes be done on material contained in fossils. The degree of decay achieved, and therefore the time elapsed since decay began (i. e. since the mineral first crystallised or since the organism died, in the case of carbonaceous material) can be calculated on the basis of the ratio of the amount of radioactive material remaining relative to the amount of the material produced by the decay. Over a limited range of relatively recent time it is also possible to measure age by counting annual rings in trees or annual laminations (warves) in lake deposits.

In the course of future redesign of Room 12, fossils of hominoids from Africa, Europe and Asia will be diplayed in the cases 3 and 4. The *Gigantopithecus* from South Asia and China with a height of about 2.80 up to 3.50 metres was the largest primate of all times. Besides thousands of single teeth only 4 lower jaws have been found so far. He is regarded as an early relative of the orang-utan together with *Sivapithecus* and *Ramapithecus*.

Africa and Europe were on the other hand the place of origin for the dryopithecines, among which the phylogenetic ancestors of gorillas and chimpanzees but also of man are to be sought.

Attempt at reconstruction of *Gigantopithecus*

13

Room 13, Fossils of Man

Before turning to the fossil remains of various primeval men, some information is necessary to understand them better. How can fossils of man be recognized as such, and how is their position in phylogenetic evolution determined?

1 To answer these questions, case 1 contains the skeletons of modern apes in typical postures in contrast to a modern human skeleton: the gibbon (*Hylobates*) swinging by its arms, the orang-utan (*Pongo*) with all four limbs used for climbing, and the gorilla (*Gorilla*) knuckle-walking, half erect but on all fours give a clear picture of the connections between bodily structure (construction) and mode of life (function).

The skeleton of a chimpanzee walking erect on its hind legs makes it clear that, although apes are capable of characteristically human gait for short periods, their skeletal-muscular system, especially in the pelvis area, is still not completely adapted to the biomechanical conditions of walking upright. In particular, the elongated dorsal bones of the pelvic girdle are still situated flat on the dorsal side so that the gluteal musculature is not spread out like a fan around the socket-joint of the hips, as it is in man. Consequently the structure of a fossil pelvic bone shows whether the creature it came from was capable of walking upright continuously, and so whether it was already a human or still an ape.

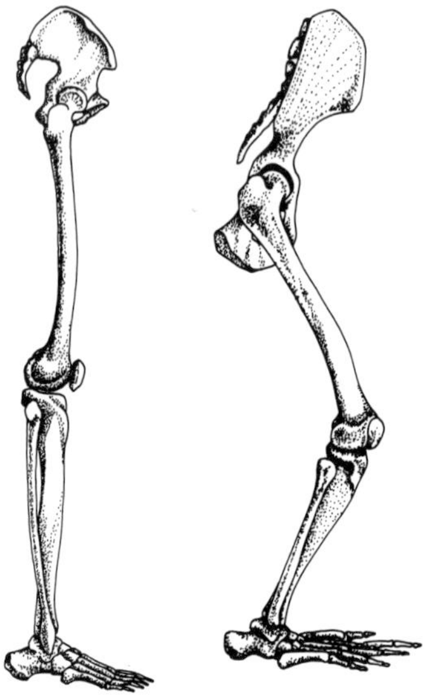

Pelvis and leg of modern man (left) and gorilla (right)

The human skeleton is divided into three large functional subunits: while the chest, arms and hands still correspond largely to those of the apes, the basin-shaped, broad-sided pelvis and the long legs are entirely adjusted to the typically upright gait of man. The almost spherical skull, on the other hand, highlights another typical human development, viz., the enormous expansion of the brain and simultaneous reduction in the size of the chewing apparatus and neck muscles.

13

Contrasting the skulls of newborn and adult humans with those of apes, and comparing them with the series *Austrolopithecus – Homo erectus – Homo sapiens*, demonstrates the interaction between the forces acting on the shape of the skull: the way the brain swells out from the inside, and the neck muscles and masseter encroach from the outside in the course of both the individual and the phylogenetic development. Phylogenetically, the human masticatory apparatus and neck muscles become smaller in the course of evolution so that the growing brain has an ever stronger effect on the skull's shape. In the development of the individual, the reverse is the case, above all in apes. When newborn, their skulls look surprizingly human, and the skulls of newborn primeval human children seem more modern than those of primeval adults. But what did our ancestors look like hundreds of thousands of years ago? *Australopithecus, Homo habilis, Homo erectus* and the Neandertal man serve as examples of how we can picture these various types of primitive man, and how such reconstructions and "portraits" are achieved.

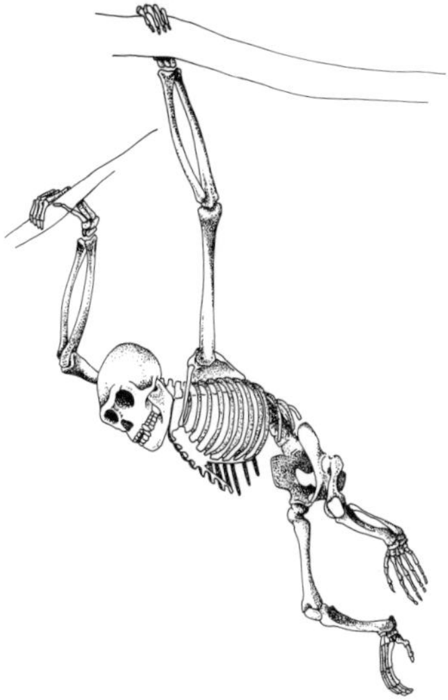
Reconstruction of *Oreopithecus*

The methods used in comparative and functional anatomy are exemplified once again by *Andrias scheuchzeri*, the "bones of a poor sinner drowned in the Deluge", and *Oreopithecus bambolii*, the skeleton of a primeval man from Italy supposed to be 8 million years old.

Localities of Fossils 2

(at present still on display as graphic No. 10) A map of the Old World provides an overview of the most important localities where human fossils have been found in the order of their discovery. The main areas so far are Central and South Europe, the Near East, South East Asia, and North, East and South Africa.

3

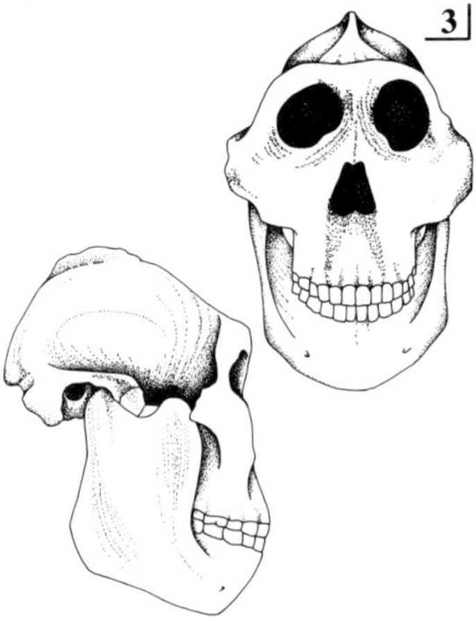

Zinjanthropus boisei

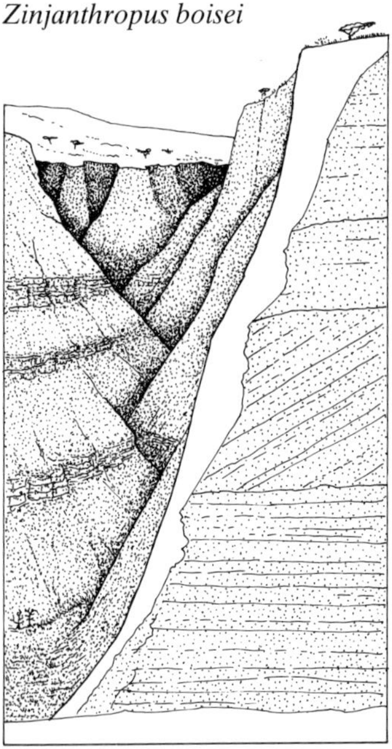

Olduvai gorge

Finds of Primeval Man in Africa I

Olduvai Gorge in Tanzania is one of the best known finding places of fossil human remains. Along with stone tools a very diversified fauna has been found in various ancient geological strata which can be dated absolutely due to inclusions of volcanic ashes (tuffs).

The lowest bed (Bed I) in the Olduvai Gorge yielded *Australopithecus robustus* (*"Zinjanthropus"* = *A. boisei*) with his massive jaws and sagittal crest, and *Homo habilis*, held to be the maker of the pebble tools found at the same site (hence the name *Homo habilis* = "the skillful man"). These hominids from Bed I are 1.7–1.8 million years old. The braincase of *Homo erectus leakeyi* (OH 9 = Olduvai hominid 9) is dated as considerably younger at 1.25 million years. It has a brain volume of about 1060 cubic centimeters and comes from the upper part of Bed II. This early *Homo erectus* probably evolved into the 150,000 year old Rhodesia man from Kabwe (formerly Broken Hill) regarded by some researchers as a form transitional to the modern *Homo sapiens*.

In contrast to the Olduvai Gorge finds, the South African hominids were found in caves without any connection to absolutely datable volcanic rocks, so their age cannot be directly determined. Therefore, the ages given are based on comparisons of the flora accompanying these finds with finds from East African sites. In addition to *Australopithecus robustus, Homo habilis* turns up again at Swartkrans in the form of *"Telanthropus capensis"*. The gracile *Australopithecus africanus* with no sagittal crest, found at Sterkfontein and Makapansgat, is estimated to be considerably older at 2.3–3 million years. The lower jaw found at Ternifine, Algeria, and named *"Atlanthropus mauretanicus"* is attributed to the North African *Homo erectus,* and the age estimated at around 700,000 years.

Finds of Primeval Man in Africa II

Recently, Africa has become the El Dorado of paleoanthropology. This is true especially for the region east of Lake Turkana (formerly Lake Rudolf) in the north of Kenya, East Africa, and also for Ethiopia.

The oldest primeval human remains found anywhere so far are those of the 3–4 million year old *Australopithecus afarensis*. While the dentition, with powerful canine teeth and the gap (diastema) between the canines and the incisors of the upper jaw, is intermediate between modern man and the chimpanzee, the structure of the knee joint shows that these early hominids were capable of walking upright. This is confirmed by the 3.7 million years old footprints discovered in petrified volcanic ash near Laetoli in Tanzania, East Africa (see case 15).

The *Homo habilis* represented by skull finds from the region east of Lake Turkana attributed to a man (No. 1470) and a woman (No. 1813), is much younger, 1.6–1.8 million years old. *Homo habilis* is intermediate in time and form between the older, delicate-boned australopithecines (*A. afarensis*: ca. 4–3 million years old; *A. africanus*: ca. 3–2 million years old), and the younger *Homo erectus* (Skull No. 3733 male, and No. 3883 female, from the region east of Lake Turkana). One of *Homo habilis'* contemporaries was *Australopithecus robustus* with huge "nutcracker" teeth. He belongs to a branch line in human evolution and became extinct 1 million years ago at the latest. Australopithecines apparently never succeeded in producing stone tools, while *Homo habilis* was able to make primitive pebble tools. *Homo erectus*, on the other hand, is held to be the maker of the hand-axes which have been found scattered over large parts of Africa, Europe and Asia. He is regarded as the ancestor of *Homo sapiens*.

4

Footprints of primeval man found near Laetoli (Tanzania)

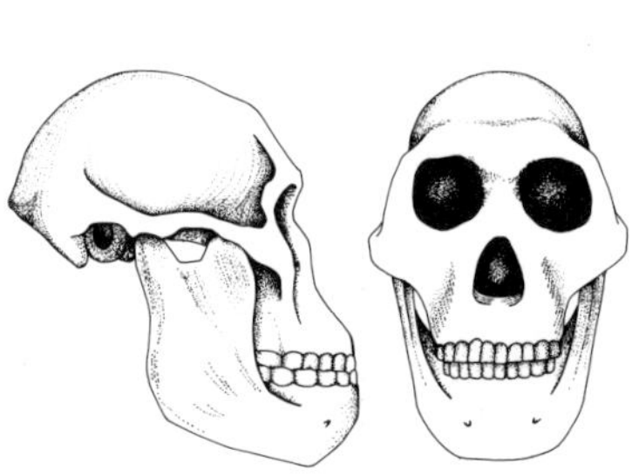

Homo habilis

5 **Finds of Primeval Man in China**

The Peking man (*Sinanthropus pekinensis* = *Homo erectus pekinensis*, ca. 500,000–350,000 years old) is somewhat younger than the *Homo erectus* from Java. He was found in a context of rather cold-loving mammal fauna (Sinosiberian faunal complex), which, however, just like the fauna accompanying the Javanese *Homo erectus* (Sinomalayan complex), can be finally traced back to Indian origins. The find of the Peking man is rather special as he, in contrast to most finds of primeval man, was found in his dwelling place, a cave, together with remains of his fire-place, his food (mainly deer and pigs), and his tools. The skull of the Lantian man (*Homo erectus lantianensis*), discovered only after the Second World War in China, is considerably older than the Peking man. His brain volume of only 780 cubic centimeters places him on about the same level of evolution as the early Javanese *Homo erectus*.

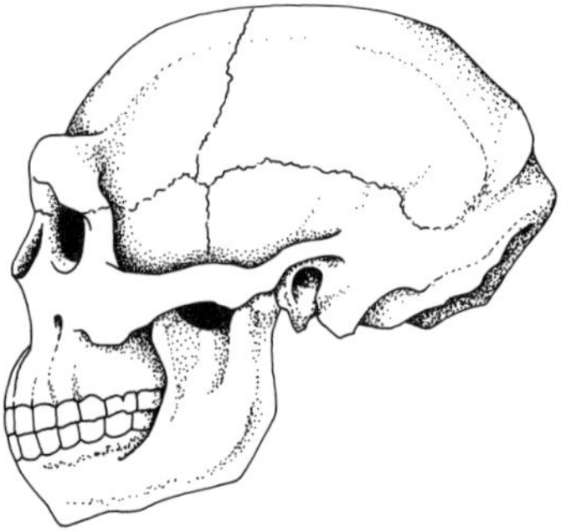

Sinanthropus (*Homo erectus pekinensis*)

6 **Finds of Primeval Man in Java**

Besides Europe and China, the island of Java, Indonesia, belongs to the areas where fossil man were found earliest. The Dutch army physician, Eugen Dubois, discovered the dome of a skull and upper thigh bones, which he named *Pithecanthropus erectus* ("the upright apeman") and described as a "man-like transitional form". However, it could not be decided that *"Pithecanthropus"* (= *Homo erectus*) really was a primitive man until 1937 when G. H. R. v. Koenigswald (who later worked at the Senckenberg Museum) discovered an almost complete top of a skull near Sangiran in Central Java. Since then, a whole series of primeval man of various ages has been found in Java, which, together with stone tools as well as animals and plants from primitive times, illustrate human evolution against a background of changing environmental conditions during the last 1.5 million years.

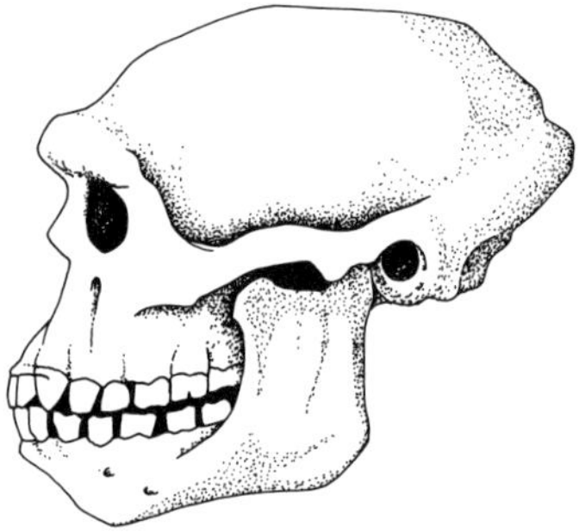

Pithecanthropus (*Homo erectus erectus*)

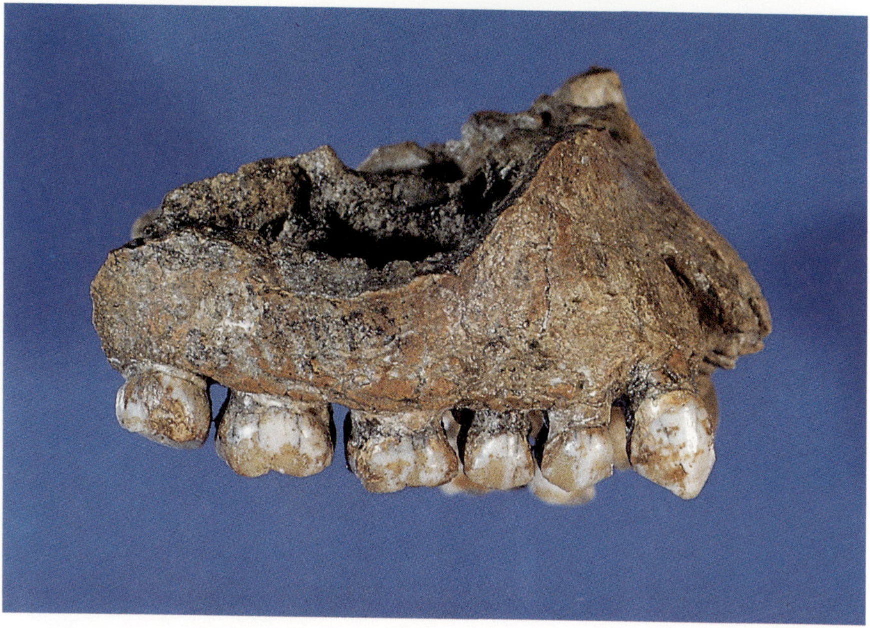

Upper jaw of Javanese *Homo erectus* (Lateral view from the right side)

Finds of Primeval Man in Europe

Finds of primeval man in Europe are divided into the Neandertal man and those which are older (Heidelberg, Tautavel, Bilzingsleben, Steinheim, Swanscombe).

European Pre-Neandertals

The Heidelberg lower jawbone belongs to a European *Pithecanthropus* form (*Homo erectus heidelbergensis*) about 500,000 years old, and so is probably the oldest European primeval man found so far. Somewhat younger, about 450,000 and 350,000 years old respectively, are the new skull finds from Tautavel, Southern France, and from Bilzingsleben, East Germany. The

7

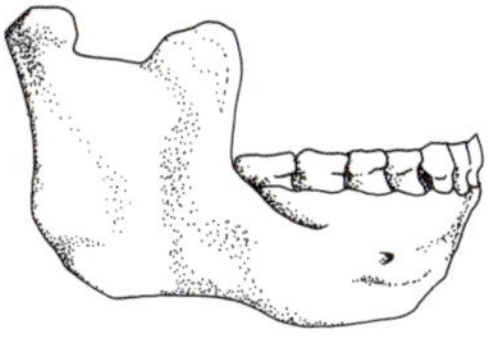

Oldest European primeval man from Mauer near Heidelberg

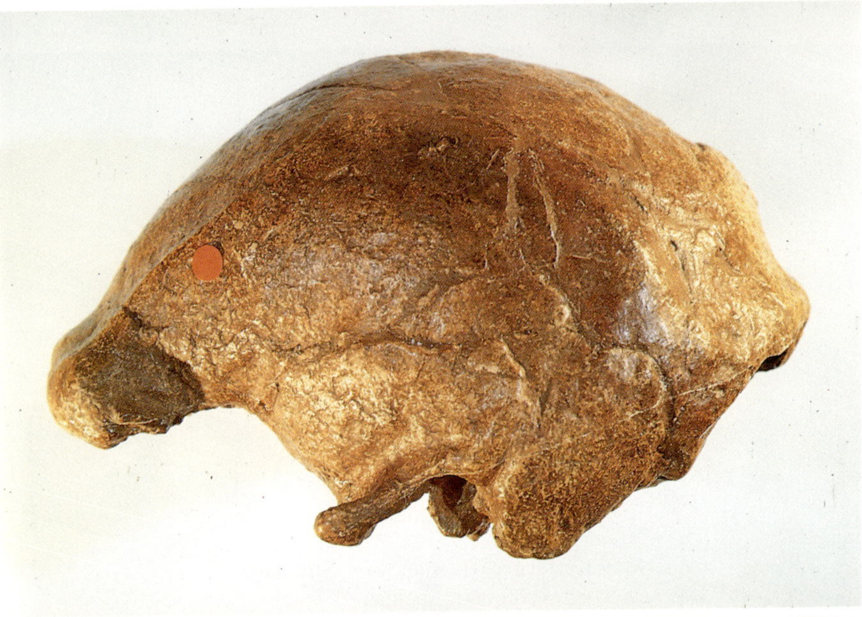

Steinheim man, aged 250,000 years, is only half as old, and might be the forerunner of the Neandertals who lived in Europe at the beginning of and during the most recent Ice Age, up to about 35,000 years ago.

Neandertal Man

The individual Neandertal finds selected are not arranged chronologically, but rather, due to the special problems involved (regionally different forms, problem of the origin of modern man), according to geographic and paleoclimatological criteria. It is now assumed that the Neandertal man was a branch line in human evolution which became extinct, while the modern European immigrated into the West European-Circummediterranean area formerly inhabited by the Neandertals from Africa via the Near East and the Balkans. Some Near Eastern cave finds of "neanderthaloid" forms of primeval man, such as the skull from Skuhl, probably represent cross forms between Neandertal man and the advancing *Homo sapiens*.

Ancestor Cult and Cannibalism

At the end of the "gallery of ancestors", modern man appears in the framework of the topic "ancestor cult and cannibalism". Tatooed and smoked Maori skulls from New Zealand are examples of ancestor worship. Skulls of killed Papuans with enlarged holes in the occiput (back of the head) or temples opened in order to extract and eat the brain bridge the gap to the past, where corresponding signs of cannibalism are known for the Peking man and the Steinheimer for example.

◄

Top: Late *Homo erectus* from Java (Ngandong man);
Bottom: Tools of Javanese *Homo erectus* from Sangiran

8

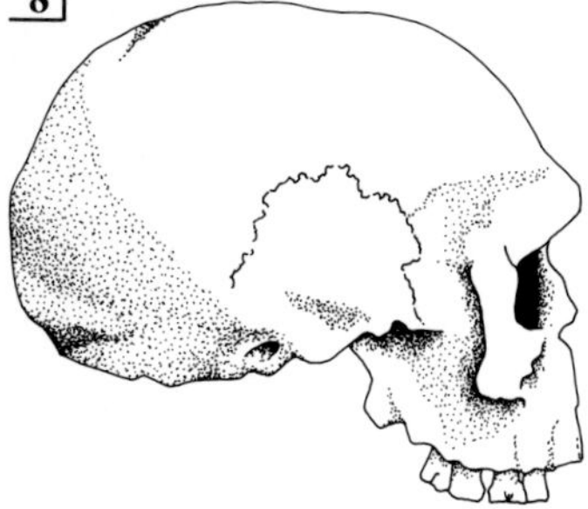

Pre-neandertal man from Steinheim

9

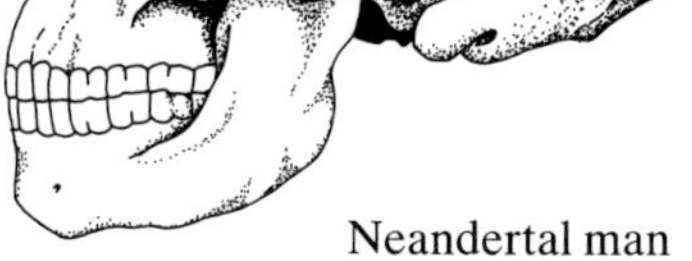

Neandertal man

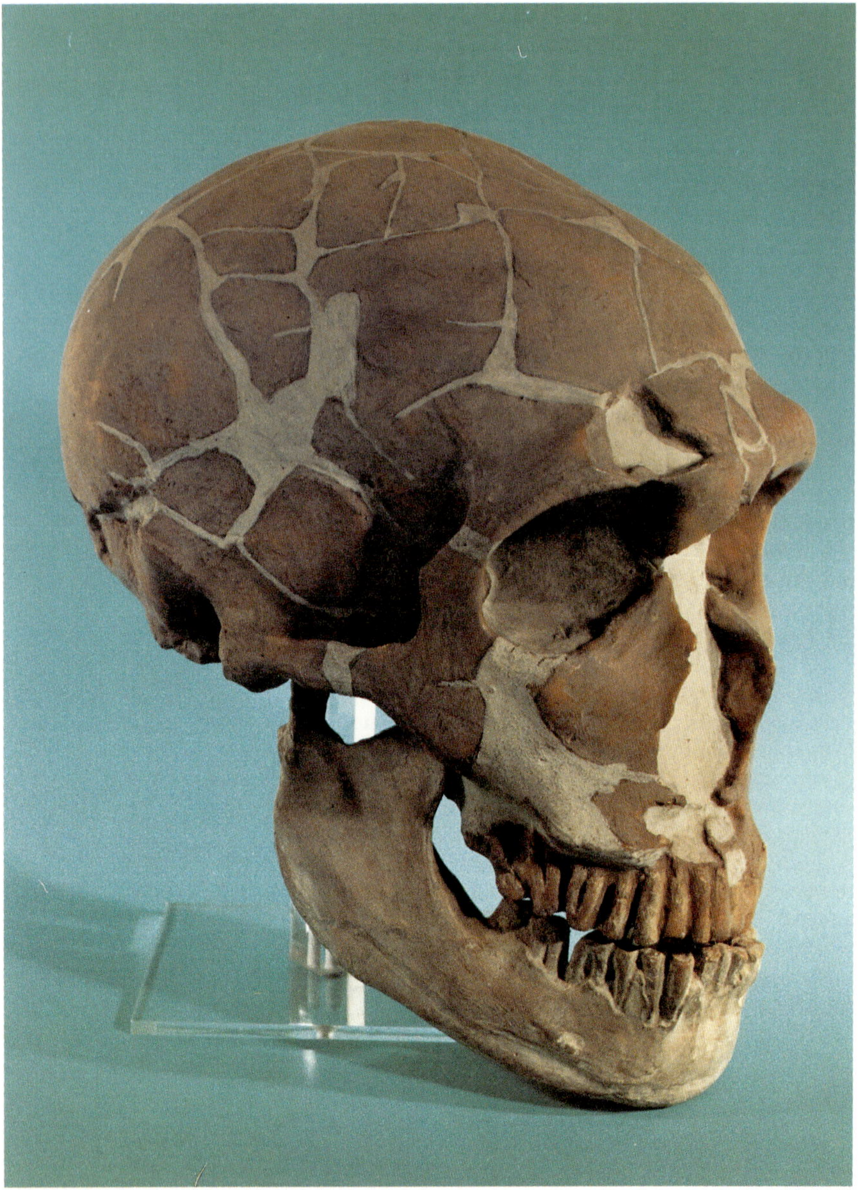

Phylogenetic Tree of Man **10**

(at present still exhibited in Room 12)
Where are the individual finds to be placed in the phylogenetic tree of man?
The decisive factor for classifying a particular find in the puzzle of the phylogenetic tree is the evolutionary stage it documents (e.g., of the brain, or of the teeth), its date, and its resemblance to other finds. The evolutionary level of individual features can only be judged in relation to certain functions, and so presupposes a functional analysis.
The evolutionary pathway to man evidently began within the dryopithecines, a group of fossil apes, which was widely spread over Africa and later Central and Southern Europe as well as South Asia about 30–8 million years ago. They were the forerunners of the Asian orang-utan and its fossil relatives *Sivapithecus, Ramapithecus,* and *Gigantopithecus. Ramapithecus,* which was regarded as the oldest hominid for a long time, is now considered to be a fossil orang-utan due to complete skulls from China as well as the microstructure of its tooth enamel. Many agreements between features, some of which are very complex (such as the structure of the chromosomes and of the body proteins), support the theory that the African apes, the gorilla and chimpanzee, together with man can be traced back to a common ancestor, which evolved only after the orang-utan group had begun a separate course of development. To date, however, there are neither any fossil proofs for this ancestor, nor any fossil records of gorillas and chimpanzees.
The earliest representative of the hominids found so far is *Australopithecus afarensis* from the

◄
Skull of Neandertal man from La Ferrassie (France)

Phylogenetic tree of man and apes ►

Venus statuette from Willendorf (Austria)

"Goddess without face", small head of a woman from Brassempouy

4 million year old Ethiopian beds. The gracile *Australopithecus africanus* is somewhat younger at 2–3 million years old, and has been found only in South Africa as yet. An early branch line in human evolution is represented by *Australopithecus robustus* from South and East Africa with a sagittal crest and massive jaws. *Homo habilis,* the oldest genuine man, and the maker of primitive pebble tools, appears contemporaneously with *Australopithecus robustus* for the first time 1.8 million years ago. The widespread group of *Homo erectus* might have evolved from him, and in turn have led to *Homo sapiens* (including the Neandertal man).
The modern *Homo sapiens,* such as appears in the form of the Cro-Magnon man alongside the last populations of Neandertals in Europe (e. g., at St. Césaire, South France) about 35–30,000 years ago, might go back to African origins (Rhodesian man, skull from Omo, Ethiopia). On the other hand the Mongolids and Australids of East and South East Asia probably represent independent regional developments.

11 **Ice Age Animals**

On the front wall of the room, there are sketches of cave paintings (not yet on display) as the background for fossil remains of the animals they depict: the tusk and vertebra of a woolly mammoth, limbs and skull of a woolly rhinoceros and an aurochs as well as the antlers of a stag. The fossils come from the last Ice Age and are more than 10,000 years old.

12 **The Beginnings of Art**

The earliest recorded artistic expressions appear in Europe together with the anatomically modern humans (*Homo sapiens sapiens*) about 30,000 years ago. Drawings scratched in rock, two-dimensional paintings done with ochre, manganese oxide and tallow, as well as bone and ivory carvings allow us today to see the animal world of the Ice Age with the eyes of our ancestors.

Youthful footprints in cave clay, and arrow and spear holes in clay statues suggest connections with puberty rites and hunting magic.
In those early times, man rarely depicted himself, and whenever he did so, it was always in a very abstract way. The so-called Venus statuettes have become famous, and are possibly to be seen in connection with the widespread mother cult of those times.

13

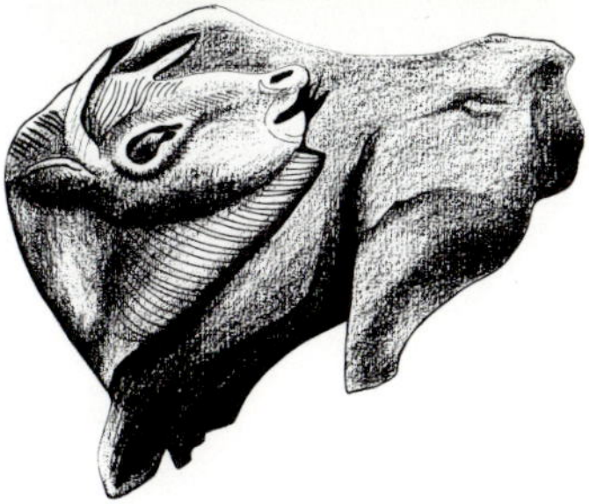

Bonecarved bison from La Madeleine (Southern France) ►

Case 13 with model of the origin of man including skull reconstructions of fossil man and skulls of all extant apes

13

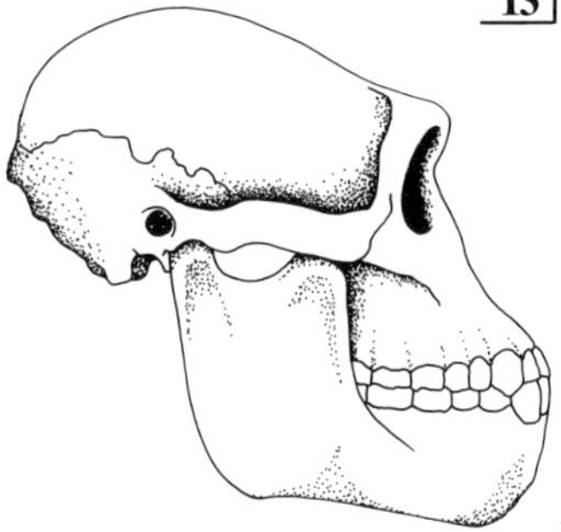

"Lucy" (*Australopithecus afarensis*), skull (reconstructed) and almost complete skeleton

13 **The Origin of Man**

While cases 3–8 on the long wall of the room contain single finds of primeval man from Africa, China, Java and Europe, an attempt to sketch a complete picture of the origin of man is made in case 13. Silhouettes depict the principal course of evolution from the still unknown, chimpanzee-like original forms via the australopithecines, *Homo habilis,* and *Homo erectus* to modern man. Skull reconstructions complete the picture of the phylogenetic development of man.

Unfortunately, the earliest phases in human evolution have not yet been proved with corresponding finds, so the skulls of modern African and Asiatic apes have been inserted as hypothetical models for this stage.

14 **The Oldest Traces of Man**

In the future, a reconstruction of the famous "Lucy", the oldest skeleton find of a fossil man so far, which comes from 3 million year old beds in Ethiopia, will round off the picture of the origin of man. This find of an almost complete *Australopithecus afarensis* made by a Franco-American expedition in 1974, together with the approximately 3.7 million year old footprints from Laetoli, Tanzania, indicates that our early ancestors were already able to walk completely upright 3–4 million years ago, while their skulls and teeth still closely resembled those of chimpanzees.

15 **The Development of Stone Tools**

Opposite the gallery of ancestors, on the glass-brick wall, the development of human intellectual and manual abilities is depicted with examples of stone tools. At first very slow, and later accelerating rapidly, the development of typical human talents can be followed from primitive pebble tools up to 2.5 million years old through to the multifarious collection of fine obsidian chippers and sharpened stone axes.

A small selection of famous human fossil finds is described in detail (in German) along with the special history of their discovery in the loose leaf folder No. 3 entitled "Fossilfunde des Menschen", which can be purchased at the admission desk, or ordered from the office of the Senckenberg Nature Research Society.

14 + 15

Room 14, Fossils from Messel (Permanent Exhibition Planned)

Temporary exhibition on Messel in Room 7, text page 58

Room 15, Earth History (Permanent Exhibition Planned)

The origin of the earth and the earth in space are subjects of astronomical research. However, events and developments on the earth after the formation of a solid crust, like origin and decline of mountain ranges, origin of continents and oceans, evolution of life, the conquest of seas, rivers, continents and the air by animals and plants are studied by the sciences of geology and paleontology. The projected geology exhibit will deal with these topics in a two-story arrangement. Thus, the structure of these will be explained as well as the endogenous processes that brought about the changes of the crust within earth history like mountain building, plate tectonics, magmatism, volcanism and earthquakes. Another priority exhibit will deal with shape and position of continents, distribution of land and sea during the various episodes of earth history (paleogeography), distribution of climate zones (paleoclimatology), and the respective state of evolution of animals and plants at the various periods. Modern tasks of geological science will be discussed (mineral resources and their role in conservation of environment).

▶
Palaeochiropteryx tupaiodon, smallest species of bats from Messel

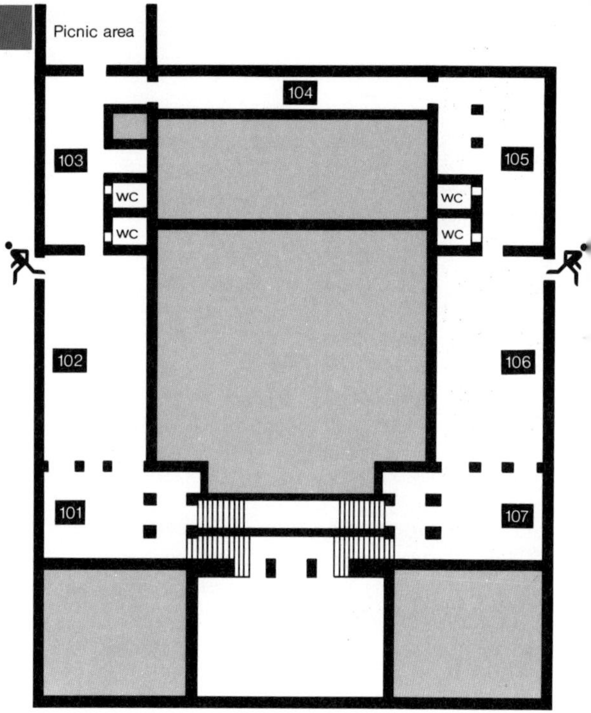

First Floor: Terrestrial Vertebrates and Plants (Planned)

Room 101–102	Mammal Gallery
Room 103	Mammals (Provisional Exhibition)
Room 104	Reptiles and Amphibians (Permanent Exhibition Planned, the Present one is Provisional) Animal Flight, Embryos, Trophies of African Hooved Animals
Room 105–107	Bird Gallery
Room 108	Paleobotany and Botany (Exhibition Planned)

Picnic Area adjacent to Room 103

Room 101–102, Mammal Gallery

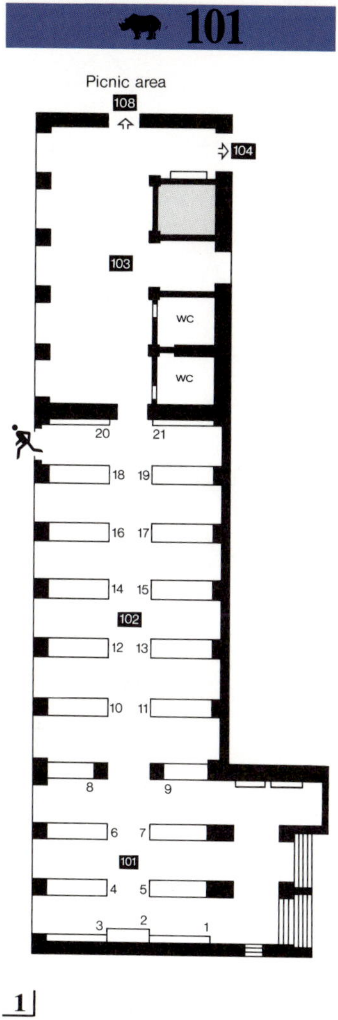

Apes

Apes are very closely related to man. One thinks of the orang-utan of Borneo and Sumatra and the gorilla and chimpanzee of the African forests. The gorilla is the largest and most powerful of the apes. It can be more than 2 metres tall and may weigh more than 300 kilograms. In contrast to the other apes it lives mainly on the ground. Beside our huge male, we display a family, acquired 1906.

The orang-utan spends almost all of its time high in the trees and rarely comes to the ground. It builds a sleeping-nest out of branches and uses it over a long period of time. The large pouch of the male, which is characterized by his cheek ridges, serves as a resonance aid in producing calls. The name orang-utan means "the man of the forest".

In spite of several points of similarity, modern man did not derive from apes of exactly the kind alive today. It seems instead that apelike ancestors gave rise to the apes on the one hand and hominids on the other. The history of man's evolution is treated in Rooms 12 and 13 on the ground floor. The skeleton of a gorilla is on display there.

Monkeys

Of the monkeys, the ones closest to the apes are the gibbons of South East Asia. They have very long arms (a gibbon's arms can span twice its height, a man's span is approximately equal to his height) which are of great use to them as they hang from trees or skillfully travel swinging through the forest. The arms are also an aid in maintaining balance when the gibbons run upright. The mandrill from tropical West Africa is notable for the strange striking colouring of its face and posterior, and the most striking feature of the gelada baboon, an inhabitant of high mountain regions in Abyssinia, is the presence of

101

Gorilla

Orang-utan

Chimpanzee

L'Hoest's monkey

hairless, blood-red patches on its breast. The great Senckenberg researcher, Dr. Eduard Rüppell, discovered the gelada baboon in the year 1832, and described it as a new species in 1835.

3 Among the most striking of the apes is the proboscis monkey of Borneo, whose nose in the adult male has the appearance of a long pounch hanging over the mouth.

In contrast to these Old World monkeys (or "narrow-nosed" apes as they are called on account of their slender nasal septum), many of the New World monkeys ("broad nosed") have a tail capable of gripping. In climbing it serves as a "fifth hand" and can even seize food. This capability is especially well developed in the spider monkeys.

Lemurs

Very similar to the monkeys in external appearance, but differing from them in dentition and
3 other features, are the lemurs (prosimians). There are no true monkeys on Madagascar and it is there that the richest variety of lemurs is found. Another creature living there is the aye-aye, whose third finger is long and slender and looks dried out. The tarsier, a small animal with very large eyes and strongly elongated tarsal bones, is an inhabitant of the Sunda Islands and the Philippines.

Carnivores

As the carnivores' principal representatives here
4 are a pair of lions. Both are Masai lions from East Africa and were captured in the wild. In contrast to the lions which one sees in zoos, wild lions are less hirsute, as is especially well seen in the relatively short mane.

Tarsius tarsius ►

101

6

Maned wolf

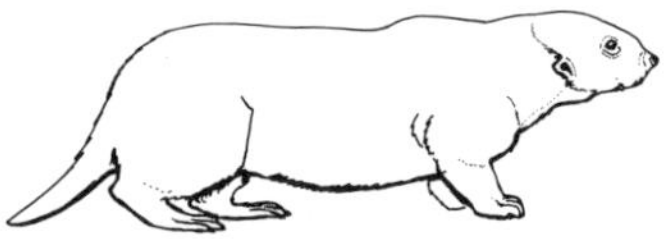

Sea otter

7

Giant panda

The maned wolf has an unusually slender body. It lives in the bushy plains of South America and feeds on small animals and fruits. The snow leopard or irbis is well adapted to life in the snow-covered high mountains of Middle Asia by its light and dense pelt. It is an excellent broad-jumber and hunts as an individual in this region. Because of its beautiful pelt it has been hunted down to a great extent. As it has become very rare it is now protected as an endangered species. A marine relative of the otter is the sea otter, which only a short time ago had become very rare because of its valuable pelt, but is now once again relatively abundant in the northern Pacific. This animal lives near the shore and feeds on seaur-chins, crabs, mollusks and fishes. The dingo lives in Australia. It probably developed from the dogs brought to that part of the world by early men moving in from the north. Today as a predactor of sheep it is therefore destroyed. The giant panda, the emblem of the World Wildlife Fund (WWF), has a striking black and white colouring. It inhabits the Hsifan mountains of Szechuan, China, and is the most habitual planteater among bears: bamboo shoots are its exclusive source of food. It has been protected by law since 1939.

Hooved Animals

The most noteworthy of the hooved animals on display is the okapi. The okapi is a short-necked giraffe fully adapted to life in the jungle. It was first discovered in the Congo (Zaïre) in 1900. Earlier, two pieces of skin from a hind leg had come to be known, but had been identified as zebra skin. For a number of years now this "forest giraffe" has successfully bred in zoos.

10 We find beside the North American bison and some dwarf antelopes and duikers a large number
11–12 of African antelopes, including the kudu, the sable antelope and the waterbuck.

15 A rather curious deer, with a long tail and its antlers apparently reversed, is the milu or Père

Quagga (*Equus quagga quagga*)

David's deer. The Jesuit priest Père David discovered it in 1865 in the Imperial game park south of Peking. The last specimen surviving in China was killed in 1900 during the Boxer Rebellion, the species, however, was rescued in captivity in European game enclosures. One hundred and fifty years ago the quagga was still abundant in Africa, but it was killed off within a few decades simply because its skin was especially suitable as material for making sacks to contain grain. Around 1878 the last free specimens of this wild

17 American bison

102

horse were killed. The last one surviving in captivity died in the Artis Zoo of Amsterdam in 1883. All of the world's museum collections shelter the remains of less than 20 quaggas. We obtained our specimen during the last century in exchange for some duplicates of Eduard Rüppell's collection. The Cape mountain Zebra, also from southern Africa, has become very rare in the wild.

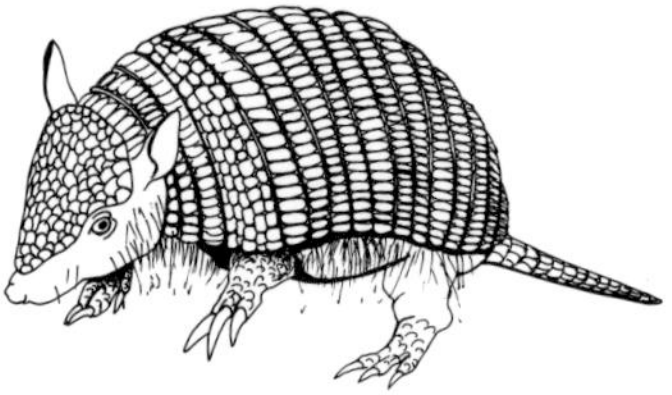

Armadillo

19 Sea Cows and Toothed Whales

The manatee on display comes from tropical West Africa. More about sea cows is found in Room 8, text on page 67. In the same case there are models of small toothed whales, which in contrast to the sea cows are derived, like the large whales, from carnivores. The porpoise is the best known whale in our waters. It is familiar to everyone who has visited the North Sea. The river dolphin from the Amazon is one of the few whale species which lives in inland waters rather than in the sea.

River dolphin

16 Pangolins and Edentates

The pangolins and ant-eaters are toothless. They live mainly on termites which they lick up with their long, worm-like tongues.

The tree sloths live exclusively in trees, hanging by their hooked claws. In some tree sloths the number of neck vertebrae is reduced to 6, in others it is increased to 8–9. Algae grow on the fur of the living animal, so that a tree sloth, hanging quietly, can look like a mossy or lichen-covered piece of a branch. The armadillos have an armour made up of ossified skin plates. They use their sharp claws to dig up the various kinds of invertebrate animals on which they feed. Noteworthy members of this group are the giant armadillo and, especially, the three-banded armadillo the mid-part of whose armour is divided into a number of girdles which can move relative to one another, allowing the animal to roll itself

Long-tailed pangolin

up into an impregnable armoured ball when it is under attack. The scales of the pangolins are made of horn, with no ossification.

Bats

18

The bats, which apart from the birds are the only active fliers among present day vertebrates, have the second largest number of species among all of the various groups of mammals (after rodents). We see here one representative of the large bats or fruit bats, the Seychelles fruit bat. Like most other fruit bats it feeds on fruits. A great deal has been learned in recent years about the ultrasonic orientation system employed by the true, small bats. In different families, the required sound waves are emitted from either the mouth or the nostrils. Among our 20 native species of bats, all of them insect-eaters, we find representatives of two families, the vespertilionid bats (the one on display is our largest indigenous form, the European mouse-eared bat) which emit their sounds from the mouth, and the horseshoe-nosed bats (the lesser horseshoe bat is shown at rest) which emit the sound waves from nostrils surrounded by a horseshoe-shaped growth of skin.

Long-tongued bat

Insectivores

19

The insectivores represent the most "primitive" group of today's living mammals. The distribution area of some families is highly limited. Thus, the tenrecs live only on Madagascar, otter shrews and golden moles exclusively in Africa. Hedge hogs, moles and shrews are, however, widespread.

Rodents

21

Rodents are herbivorous to omnivorous. They possess a pair of incisive teeth with permanent growth in both the upper and lower jaw. The overall length of animals of this group varies between 12 cm (dwarfed mouse) and more than 100 cm (capybara). The capybara lives near

ponds and rivers in South American forests with dense undergrowth. It is an excellent swimmer and diver. Its main enemy is man who kills it because of its meat.
Local representatives are on display in small dioramas.

18 **Marsupials**

In the marsupials, primitive mammals from Australia and America, the young are born in an undeveloped state. At birth, the young of a giant kangaroo is only as long as a finger joint. After birth the young make their own way into the pouch, a fold of skin on the belly of the female where they find the teats and it is there that they continue their growth. There are several different kinds of marsupials, all of them externally similar to other mammals. In addition to the large plant-eating kangaroos and the exclusively arboreal marsupial bears (e. g. the koala) there are also predators like the Tasmanian devil and the now probably extinct Tasmanian wolf. The opossum,
20 the best known marsupial in the New World, is likewise a predatory marsupial. Other marsupials have the forms of flying squirrels, rats, mice, shrews and even moles. A selection of particularly characteristic small forms is brought together in our exhibition.

Kangaroo

20 **Monotremes**

Standing in contrast to all other mammals, which bring live young into the world, are the monotremes which like birds and most reptiles lay eggs. Members of this group are the duck-billed platypus and the two genera of echidnas. These toothless animals, provided with beak-like specialized jaws, are the most primitive mammals alive at the present time. The aquatic platypus

Echnida

Koala (*Phascolarctos cinereus*) ▶

uses its bill (which is not comparable in anatomical structure with that of a bird) to search for its food – crabs, worms and small fish – by rooting around in the mud at the bottom like a duck. The eggs are laid in nesting holes at the edge of the water. This animal's home is in Australia. The echidna, by contrast, lives on dry land. Its egg is incubated in a fold of skin on the belly of the mother. The echidna's home region is Australia and New Guinea.

Room 103, Mammals (Extension of Mammal Gallery; at Present Provisional)

This room will hold part of the new mammal exhibition and now houses a provisional exhibition of skeletons and mounted specimens of larger mammals.

On the wall there is a skull of an Indian and an African elephant. A complete skeleton of an Indian elephant and a dermoplastic of an African one are on exhibition in Room 10 on the ground floor.

Among the large mammals in this room the giant "Irish elk" is already extinct. It was abundant in Europe during the Ice Age and probably still survived to be a contemporary of the Teutons. Like all deer, it shed its ponderous antlers every year. It inhabited open marshy and boggy areas. The skeleton on display comes from Pleistocene deposits of a bog in Ireland.

The Alaskan moose or elk is the largest of the modern deer. Its antlers span up to 2 metres. It lives in the northwestern part of North America.

The "rhinoceros group" is represented by skeletons and dermoplastic of different species of rhinoceroses and a tapir. The Great Indian rhinoceros is confined today to Nepal, Assam and Bhutan. Only a few hundred animals of this species exist, as it was – and is still today – hunted

Wide-mouthed rhinoceros

for its horn, which is believed to possess medicinal properties. The wide-mouthed rhinoceros has two horns. It feeds on grass and was distributed over large parts of Africa. Today it lives in the Sudan, perhaps still in Uganda, in adjacent areas of Zaïre, in South African game reserves, and – introduced – in the Krüger National Park.
Skeletons of an okapi, a Grevy's zebra and a Malayan tapir are added for comparison.

104

Room 104, Reptiles and Amphibians (Permanent Exhibition Planned, the Present one is Provisional); Animal Flight, Embryos, Trophies of African Hooved Animals.

In this display room, presently only accessible as a narrow passage way, a provisional exhibit of amphibians and reptiles, parts of a former special exhibition on the flight of animals, trophies of African hooved animals and embryos can be studied.

Animal Flight

The exhibition on "Animal Flight" records how representatives of several different groups of animals have conquered the air.

There were already insects with fully developed wings in Palaeozoic era (in the Upper Carboniferous, approximately 325 million years ago). We are aware that animal life today includes all sorts of insect groups which display a wide variety of modes of flight.

Among the vertebrates it was probably reptiles which were the first to become capable of flight at about the beginning of the Mesozoic era. The flying reptiles of Jurassic time (see Room 9) were already astonishingly well adapted to life in the air. Like the modern bats, their aerofoil surface was an extension of their skin, not stretched between the fingers however, but instead spread out between the much elongated 4th finger and the body. The size of the flying reptiles ranged from sparrow size up to that of the giants of Cretaceous time which had wingspans measuring as much as 8 metres like *Pteranodon.* In Mesozoic era the birds developed into excellent fliers, too . Here we display in addition to the skeletons of *Archaeopteryx* as the most primitive bird (see room 8), a "stuffed" reconstruction in which starling feathers have been used.

Insects, flying reptiles, bats and birds have all adopted the rowing mode of flight in which forward motion results from up and down beats of the wings. Vibratory and humming modes of flight have also been developed. Many birds are able to glide, holding their wings still and exploit uplifting forces in the air. A number of animals can launch themselves into brief acts of gliding, in which they may nevertheless cover impressive distances. Flying fishes, for example, thrust themselves out of the sea and, using their wing-like elongated pectoral fins, they can glide over the water for distances of several hundred metres. Fishes of this kind have chanced on the biplane principle, with pelvic fins also elongated and serving as supplementary airfoils. The flying squid has developed rocket thrust. These squids shoot into the air, driven by reaction to a backward-directed jet of water. At the peak of their trajectory they spread out lateral membrans which serve as airfoils and stabilizers.

On land there are animals which can descend from heights, e. g. the tops of trees, using parachute like expedients in order to float down. Amphibians which do this are the Javanese flying frogs which have large webs of skin spread out between their long toes. Among the flying mammals the gliding possums, flying squirrels and flying lemurs use flaps of skin spread out between the fore and hind legs, and in some cases between the hing legs and tail also, in order to glide. Among reptiles, the Javanese flying dragon, is a glider. Its skin surfaces can be spread out by the action of elongated spreading ribs.

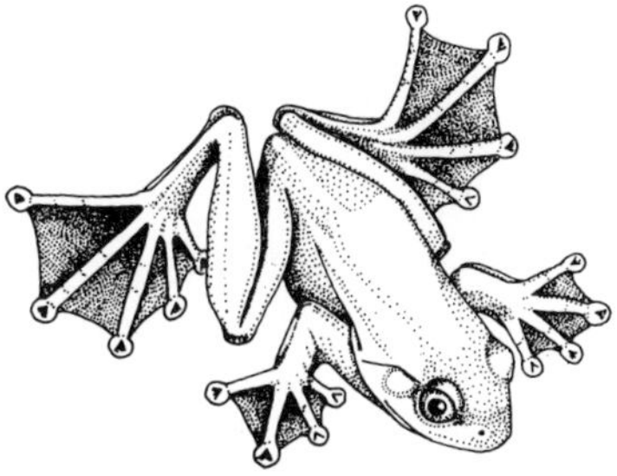

Flying frog

Embryos

In the passage leading to Room 103 two cases show mammal embryos. It is to be understood that the embryos are at all stages contained within a fluid-filled sac which protects them against any external pressure and also allows them freedom of movement. Exercise is already necessary

104

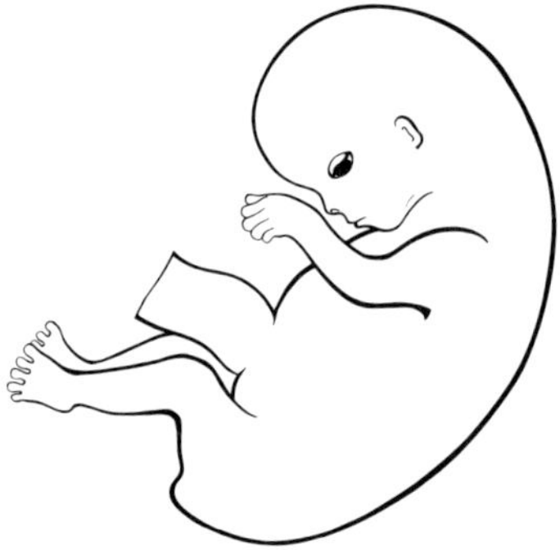

Embryo of man, 4 weeks old

at the embryo stage in order that joints and connective tissues may develop normally and establish their proper functions. This, too, deserves note: the human embryo in its 2nd and 3rd month of development already approaches human form. All of the organs are present and the differences between the embryo and the child at birth are now mainly matters of proportion. The head appears unusually large and the arms and legs are proportionately small.

The envelope containing the embryo is provided with tasselate structures which link into the blood vessels of the embryo. The whole sac-like structure, complete with tassels, is called the placenta. The tassels are inserted into the wall of the womb and so lie immediately adjacent to the mother's blood vessels. By this means, which involves no interfusion of the mother's blood vessels and those of the embryo, the growing being acquires a connection leading through its umbilicus into the circulatory system of the mother. The bodily needs of the embryo are provided for, in that nutriment and oxygen passes from the mother's blood into that of the developing child and the embryo's metabolic waste takes the opposite course. After birth, the placenta is no longer needed. The process of birth usually begins with the bursting of the fluid-filled sac. The child is born and the placenta is subsequently ejected as the afterbirth. At the moment when the child takes its first breath of air, it is no longer dependent on the mother's circulation system and the umbilical cord can be severed and detached.

In most of the mammals, the placenta, although still a structure essential to mammalian function, is not so well developed as in man. All mammalian embryos do of course lie in a sac similar to that which contains the human embryo, and enveloping fluid provides them with the same protection against damage and the same opportunity of exercise; but the tassel structures which provide the link with the body of the mother animal are arranged differently. In hooved animals the placenta can in most cases be described as patchy (Placenta cotyledonaria) and in carnivores there is a girdle of tasselate structures running round the embryo's envelope (Placenta zonaria). The embryo of the hare on the other hand has a disc-shaped placenta, which does not, however, surround the whole of the embryo's sac (Placenta discoidale). The placenta of man and of many monkeys too is discoidal but it covers, especially in man, the whole surface of the embryonal sac. The most primitive arrangements for embryonal development found in mammals are seen in the marsupials, whose young, born very early while they are still genuinely embryonic, take up a position at the outlet of the milk ducts within the mother animal's pouch.

Trophies of African Hooved Animals

The heads of African hooved animals, prepared the way they were collected by hunters of large animals as hunting trophies, show the variability of the frontal weapons of these animals.
Most of the species exhibited here are now endangered by overkill and expansion of man.

Lesser kudu

Reptiles (see also Room 105)

As we trace our way through the evolutionary sequence of vertebrates, the reptiles present the first examples of vertebrates which can spend all of their lives on land. While the body temperature of the higher vertebrates, the mammals and birds, is independent of the environment, all reptiles and amphibians have variable body temperatures corresponding to that of their surroundings.
Turtles, crocodiles, lizards, snakes and tuataras belong to the reptiles of today. These animals are very diverse in external appearance, but an array of common characteristics points to their close relationship. The embryological development of the reptiles, like that of the birds and mammals,

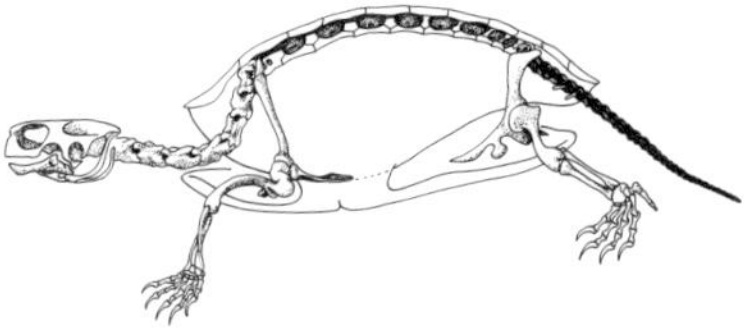

Skeleton of a tortoise

Box tortoise

always takes place outside the water and for this reason the embryo is protected by special coverings and a calcareous eggshell to prevent the risk of drying out. The skin of all reptiles is covered by horny scales or plates which help to reduce evaporation. The hatched young ones resemble their parents, there is no larval stage.

Turtles

The turtles are a deviant group. Their basic design is nevertheless in evidence quite early in the evolutionary development of the whole reptilian group. Armour composed of bony plates and covered with thin horny scales encloses almost the entire body. When threatened, most of the turtles can withdraw their head, limbs and tail into the shelter of this armour. Land tortoises which have a heavy, highly vaulted shell, predominantly feed on plants, which were cut up and chopped with horny jaws. The most familiar is the Greek tortoise from southeastern Europe, which was very often kept as a pet. Tortoises attains a much more significant size in Africa, e. g. the leopard tortoise and the African spurred tortoise. The tortoise from Aldabra Island and the Seychelles in der Indian Ocean (see Room 8) and the elephant tortoise of the Galapagos Islands in the Pacific Ocean become even larger.
The existence of all turtles and tortoises is threatened, either because their habitats are being reduced, or because they are directly hunted for their meat, tortoiseshell, or eggs.
The European swamp turtle is the only representative of this reptile group in our native fauna. In the nature reserve “Enkheimer Ried” within the boundary of the City of Frankfurt it may be watched at sunny weather. The largest and heaviest fresh water turtle is the alligator snapping turtle from the southeastern United States. With its razor-sharp jaws, it chops up its prey, consisting of small vertebrates. The New Guinea river turtle from New Guinea and North Au-

● 104

stralia shows suggestions of the soft-shelled turtles in its anatomical characteristics and can be viewed as a connecting link. The peculiar snake-necked turtles from the southern hemisphere cannot retract their unusually long neck under their armour but bends it sideways when endangered.
The marine turtles are at present exhibited in Room 8 and therefore mentioned on page 69.

Crocodiles

The crocodiles are the only direct relatives of dinosaurs which did not die out at the end of Cretaceous time. They are, however, only sparingly represented at the present day. There are, in all, 23 living species. They live in the rivers and lakes of the tropic and subtropic zones and in some cases reach huge proportions, with lengths of as much as 10 m. The powerful tail provides the drive for these strong swimmers. They can, however, on land, too, move their raised bodies at astonishing speed. The long jaws bear numerous pointed teeth which are employed to seize and hold the animals on which the crocodiles prey. Depending on their size, crocodiles will take fish, amphibians, swimming birds and mammals, but always swallow them in the water. Crocodiles have been, and are still today, hunted for their skins and because of this they have disappeared or have become very rare in many regions of the tropics. The South American black caiman, displayed here, is one of the crocodile species which is threatened with extinction.

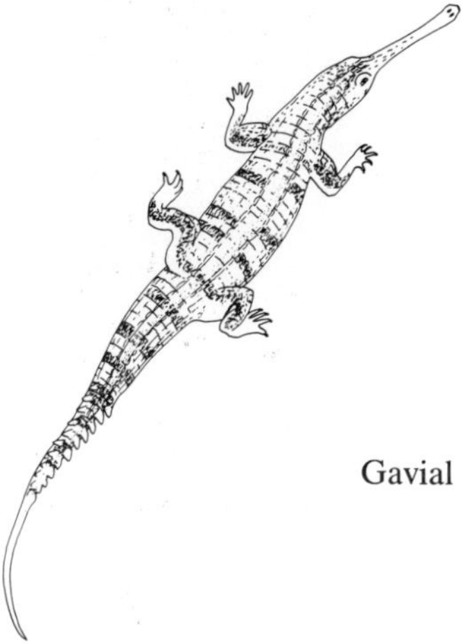

Gavial

Tuatara

A primitive reptile alive today is the tuatara, found on some small islands and cliffs off New Zealand. It has a double bridge at its temples and therefore represents an intermediate form leading to those lizards and snakes in which this bony bridging is much reduced or fully eliminated. Obviously the tuatara has been able to survive on

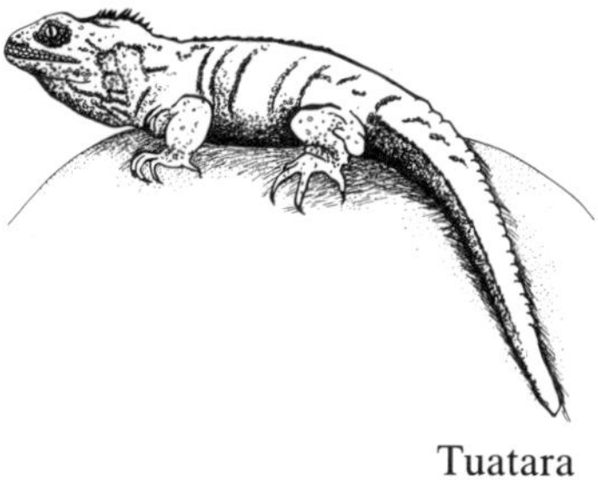

Tuatara

Giant girdled lizard (*Cordylus giganteus*)

Gila monster (*Heloderma suspectum*) ►

New Zealand only because there it is not exposed to competition from "modern" reptiles and mammals. On these islands the tuatara lives in holes which it digs itself and which it often shares with petrels. It is plainly well suited to the cool, equable climate there, with average annual temperatures around 13 °C, which is very unusual for reptiles.

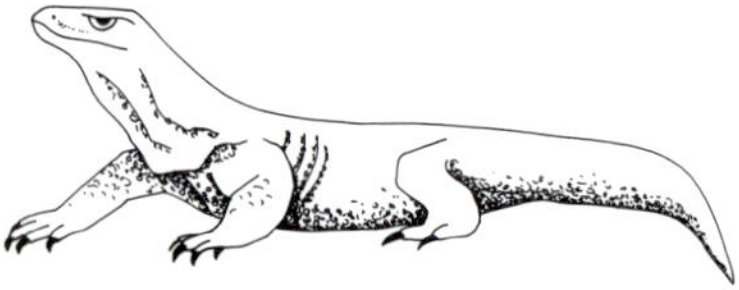

Komodo dragon

Lizards

The mostly small, quite varied reptiles of which the European sand lizard and the leg-less slowworm serve as local examples, belong among the lizards. The largest lizards are the monitor lizards. We display the 3 meter long Komodo

Moloch (*Moloch horridus*)

dragon and the water monitor which remains thinner. The slender kinetic skull of the monitor lizards and their long drawn out trunk and tail suggest to students of this group that they must be related to the forerunners of the snakes.

The iguanas are the most widespread lizards of the New World. The green iguana, which reaches a length of up to 1.5 metres, is one of the best known species. In contrast to the monitor lizards it feeds mostly on plants. Its relative, the peculiar marine iguana, which Charles Darwin observed on the Galapagos Islands, swims and dives in the sea and feeds on algae, weed and marine invertebrates.

Three Australian species representing the Old World agamids are on display. The waterdragon or water devil can raise itself up for rapid movement and can run on its hind legs. The raised neck-frill of the frilled lizard produces a dramatic addition to the outline of the animal and makes a threatening effect on any attacker. The moloch is protected by a number of spiny plates, but is – its name notwithstanding – a harmless small lizard which feeds exclusively on certain particular species of ants and for that reason cannot be kept in zoos. In spite of its bright colour it is very hard to distinguish it from the red sands of the Australian deserts.

The gila monsters from the southern part of North America include the only two poisonous species of lizard. In contrast to snakes, their venom fangs are in the lower jaw. The poison, like the venom of many snakes, affects the nervous system of the victim and it can even be fatal to humans. Impressive figures among the tree-dwelling lizards are the chameleons, which are widely known because of their ability to change colour. Their distinctive features are their prehensile claws, the gripping tail, the spherical eyes which can move independently of one another and the catapult tongue. The tongue has a clubbed end and can be shot out to distances of the order of the body length in order to trap insects and bring them back between the jaws to be swallowed.

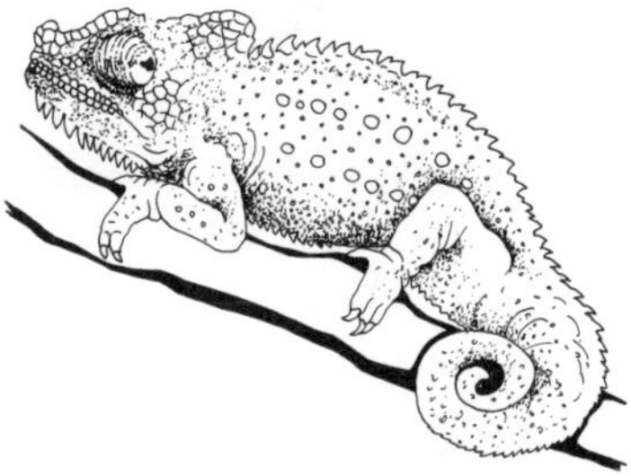

Two-lined chameleon

Snakes

Snakes are close relatives of the lizards. They are easily distinguished because they have lost their limbs, the pectoral and pelvic girdles are atrophied, their eyes shine through the lids transparent overgrowths, and also because of the extreme flexibility of their jaw bones.

They are only sparsely represented among our native faunas. The best known among them is the ringed snake, which can occasionally be encoun-

Sand boa

Ringed snake

Common viper

tered among rich plant growth bordering ponds and lakes. It seeks its food, frogs and fish, in the water mostly. The splendid Aesculapian snake, which like the ringed snake is entirely harmless, is found at only a few, climatically favourable localities in Germany. The city of Schlangenbad in the Taunus is named after this snake. The Aesculapian snake is an avid consumer of mice and has thus an important part to play in local ecological relationships. A more widespread form is the aggressive, but again non-poisonous, smooth snake, which because of its pattern is often mistaken for the common viper and consequently killed. The common viper is absent from Frankfurt and its surrounding region and from the Taunus Mountains. The nearest area in which it occurs is in the northern part of the Spessart Mountains. Of the foreign snakes, the giant snakes, at least, are fairly generally known, and a prime example is the boa constrictor. The giant snakes, not all species of which are of large size, retain vestiges of the pelvic girdle and other primitive features, and for this reason they are placed, in systematics, near to the origin of the snake group. They are entirely non-poisonous and kill their prey by winding themselves round it and strangling it (see Anaconda in Room 8). The skeleton of the reticulate python on display gives us a good insight into the internal structure of the snakes and we note especially the individually moveable skull bones and jaw bones and the regularity of the 377 vertebrae, each with its pair of ribs.

Although only about a fifth of all snakes have venom glands (specialized developments of their salivary glands) and therefore belong among the poisonous snakes, a suspicion of danger is attached to all snakes. Poisonous snakes – at most the larger species's bite can be dangerous to man – are distinguished by the fact that in the upper jaw they have at least one pair of venom fangs which have connections to the poison

glands and each of which has a slit or a canal for injection of the venom.
The whip snake, a representative of the tree snakes, is on display. Their horizontal pupils and their narrowed nasal region allow them to fix both eyes on a victim and so to make an exact judgement of its distance. There are no exterior characteristics which distinguish poisonous from non-poisonous snakes. The case is often quite the opposite – some harmless South American snakes have exactly the same black-yellow-red colouring that one finds in the coral snakes, a genus of poisonous snakes with an abundance of species. Australia has a large variety of poisonous snakes and one of the most feared among them is the black snake.
The threatening attitude adopted by the spectacled cobras is impressive – the head and the anterior part of the body are lifted up and the neck-ribs spread out so that the spectacle-like signs become visible. The potential antagonist is carefully watched, but the cobra will not bite provided that its opponent keeps its distance. "Snake charmers" make use of this trick of behaviour even when dealing with snakes which are dangerous to man and whose fangs have not been tampered with in any way. Some spectacled cobras and the spitting cobra of Africa can shoot their venom into the eyes of an attacker with such accuracy that it will immediately be aware of the irritant effect on its mucous membranes and break off the attack. The largest cobra and the longest poisonous snake of all is the king cobra from Southeast Asia which can attain a length of 4.57 metres. The specimen on exhibit is almost 4 metres long. The king cobra feeds almost exclusively on other snakes.
In contrast to the elapid snakes, the vipers are plump in build and are mostly lurking hunters operating at ground level. In the vipers and pit vipers each upper jaw bone bears solely the very long poisonous fang (and its replacements) which

● 104

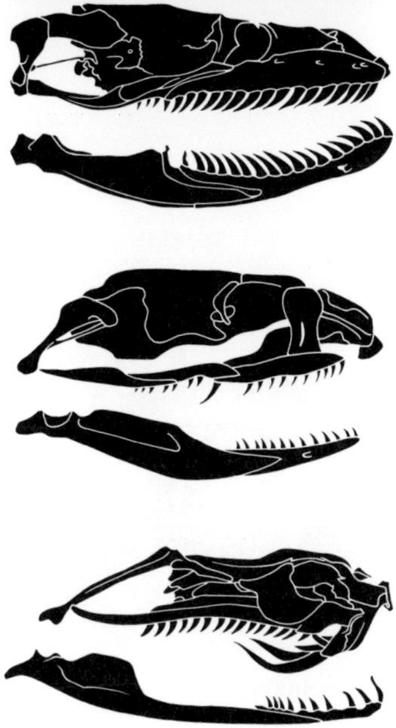

Skulls of snakes (from top): Giant snake, Boigine viper, puff adder

Spectacled cobra

Rattle snake

Japanese giant salamander

has a closed tube for supply of the venom, opening to the exterior just short of the tip of the fang. Since the jaw bone can be rotated about its transverse axis the fang is capable of being erected when the mouth is opening and can function at the instant of biting like a needle delivering an injection. One of the African vipers is the short, fat Gaboon viper. One group of vipers has a depression situated between the eyes and the nostrils called a pit-organ. This is a delicate sensor of temperature differences and can be used to trace warm-blooded prey. Beside a South American lance-headed snake we see a bushmaster, the largest of the pit vipers.

Many vipers, when excited, set the tips of their tails into quivering motion and so produce a rustling sound among the fallen leaves on the forest floor. The rattle snake, when moulting, retain one dry ring of skin at the end of the tail (which adds up to 8 rings in the course of the years) and using this "rattle" can make a loud announcement of its presence. This rattling has been interpreted as a warning signal directed at large mammals which could be dangerous to this snake.

Amphibians

The amphibians are plainly incompletely equipped for life on dry land. Their skin is moist and abundantly supplied with glands, and the mucous secretions are only moderate protection against desiccation. There are no horny plates or scales to help. For reproduction most of them must return to the water where the eggs are laid and the larval period is spent. Only when metamorphosis is over does the gill-bearing tadpole became a lung-breathing amphibian. There is, however, a number of species without a tadpole stage. Among modern animals we count among the amphibians the newts and the salamander, the well-known frogs and toads, and a group present only in tropic regions, lacking limbs and worm-like in form, the caecilians. Unlike reptiles, the newts and the salamanders are distributed with many species in the temperate regions of the northern hemisphere, whereas the frogs and the caecilians attain their greatest variety in the tropical zone.

From the approximately 6,000 known species of amphibians we select for display only some species of our native fauna and the largest one of each of the salamanders and the frogs.

The Japanese giant salamander is an inhabitant of cool mountain streams. At 1.60 metres long it has what are for an amphibian huge proportions. The giant salamanders are restricted to eastern Asia nowadays, but Tertiary deposits provide evidence that they were distributed worldwide at that time. One of the largest frog-like amphibians is the West African Goliath frog which can weigh as much as 2 kg. It lives in rivers and streams rich in oxygen.

Native Amphibians

Our indigenous amphibia are represented by some salamanders, frogs and toads. These animals are little noticed because they lead nocturnal lives. However, in the spring mating season the males call to draw attention to themselves and to attract females into their spawning waters. The compulsive migrations of toads toward the ponds where they will lay their trails of eggs often brings them to attempt to cross busy roads where the traffic can cause mass fatalities. But our native amphibia are much more gravely threatened by pollution, or by filling in the small bodies of standing water in which their eggs are laid and their larvae grow. One of the most endangered domestic amphibians is the tree toad. Aware of this threat to the amphibians, an attempt to conserve and protect this animal by cultivation of new wet areas for spawning.

Midwife toad

Tree toad

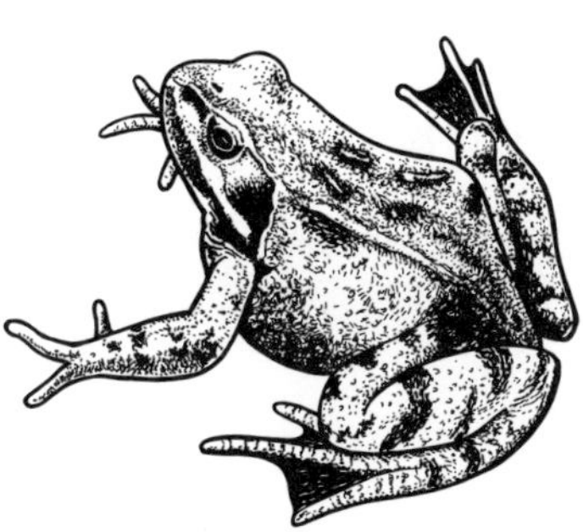

Common frog

107

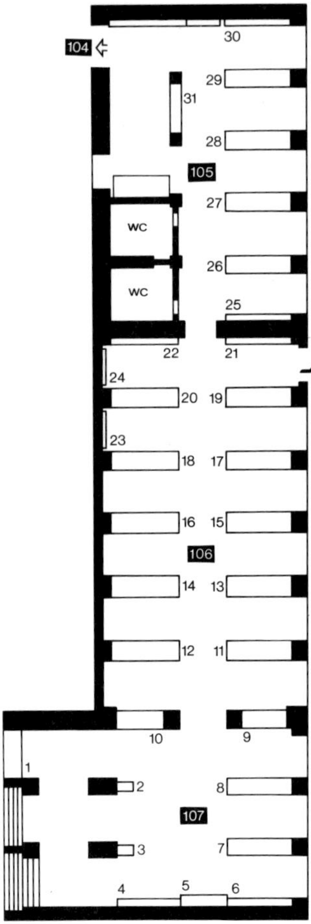

Room 105–107, Bird Gallery

Our exhibition attempts a comprehensive review of all of the orders and families of present-day birds. We can show only a small proportion of the almost 8,600 known species and we have selected the ones which deserve special interest because of their appearance, way of life, or behaviour. Some are therefore shown at their nesting-places, with their nests. Others are mounted in particularly striking attitudes.

1 **Moas**

The best starting point for a tour is at the case beside the stairs in Room 107. In it are three skeletons of moas, huge ostrich-like birds. In the middle stands *Dinornis*, the largest moa, whose height reached 3.30 metres. The ponderous *Pachyornis* is on the left and on the right is *Emeus* which most closely resembles the present day emus of Australia and the rheas of South America. Like these ostriches, the moas had three toes, whereas the African ostrich has only two. All of the moas lived in New Zealand. They met their extinction in historic times, the last ones dying about 200–300 years ago, as abundant remnants of them prove – skin, feathers, bones found in caves. Since 1870 more and more localities have been found where remains of these giant ostriches lie preserved and finds of complete skeletons have been reported quite recently. Like New Zealand with its moas, the island of Madagascar also had giant cursorial birds which survived into historic times. The Madagascar form, called *Aepyornis*, however, is not closely related to the moa. An *Aepyornis* egg is on display in the ostrich case.

Great Auk and Dodo

Moving in that direction we encounter on the pillars two birds killed off by man: the great auk and the dodo. Remains of these are among the most highly prized items in natural history collections. The Great Auk is a relative of the auks and guillemots. Up to 150–200 years ago it was widespread and abundant along the north Atlantic coasts. The Great Auk could not fly. Like the penguin, it used its short wings as paddles in swimming. Today there are only about eighty skins and a few skeletons in museums.

The Dodo was a heavy, flightless bird which lived on the island of Mauritius in the Indian Ocean. Soon after the discovery of the island, some 250 years ago, the Dodo was killed off by sailors who saw the defenceless birds as an easy source of meat. Although at least two live specimens were brought to Europe in the 17th century, neither skins nor eggs have been preserved. All that remains is a few skeletons, one head and two feet.

Ostriches

Three cases are given over to ostriches. The one on the left shows different forms of the cassowary, on the right there are rheas, emus and kiwis, and the middle one shows African ostriches. A special feature of the kiwi is its highly efficient sense of smell (an exceptional development among birds) which is a feature linked to its nocturnal habits. It is cloaked in a cover consisting of narrow, hair-like, loosely hanging feathers which entirely conceal the short stumpy wings. Under the ostriches the eggs of five different kinds of birds are laid out for comparison of sizes: the egg of the extinct giant ostrich *Aepyornis* corresponds in volume to 6 African ostrich eggs, 9 cassowary eggs, 148 eggs of the domestic fowl or 50,000 humming bird eggs.

Penguins

Many birds live at sea. Among them are the penguins, inhabitants of the southern hemisphere, especially the Antarctic region. They follow the cold water currents northward as far as

2

3

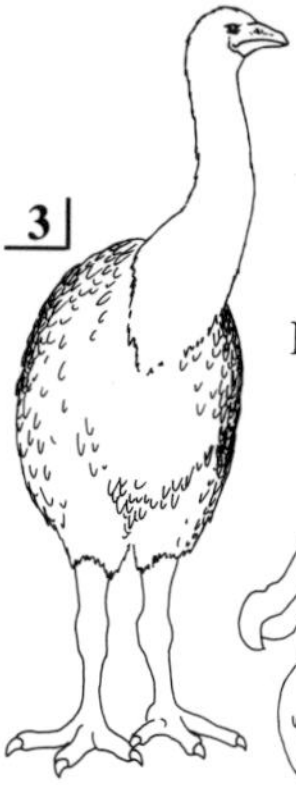

Moa

4–6

Dodo

Ostrich

7

Kiwi

southern Africa (Walvis Bay) and up the west coast of the Americas as far as the Galapagos Islands. They cannot fly, their wings are short and paddle-shaped, the feathers transformed into minute, almost scale-like growths. They are extremly social creatures and there are colonies numbering hundreds of thousands of these birds. The large species usually lay only one egg which is incubated resting on top of the feet and covered by a fold of the belly skin. The smaller species lay 2 eggs in the shallow depressions these birds use as nests.

Albatrosses

Albatrosses are especially well adapted to a life spent above the oceans. They can make extraordinarily good use of changing currents of air and are the only birds to employ dynamic gliding techniques. Thus they can follow a ship for days, picking up food material thrown overboard. Equally masterful in the airspace above the oceans are the procellariids and storm petrels. The shearwaters, members of the same general group, are more strongly adapted to diving.

Totipalmate Swimmers

Members of the group of totipalmate swimmers, the frigate birds and the tropic birds, are found in tropic waters. Among the cormorants some species have been trained for fishing in China: These birds dive for fish in their accustomed way, but a ring placed around their neck prevents them from swallowing their catch, which their masters recover from their throat pouch. Pelicans, in spite of their plump appearance, are splendid fliers and skilful fishers. They use their huge beak and extensible throat pouch like a dipnet. One can observe large flocks of them feeding together in shallow waters.

◄
Great auk (*Pinguinus impennis*)

Emperor penguin

7

Albatros

8

American white pelican

107

Stork-like Birds and Flamingos

The group of stork-like birds includes, in addition to the storks themselves, the herons, the whale-headed storks, the spoonbills and the ibises. Among the herons we are immediately struck by
9 the appearance of the Goliath heron from Africa, 1.5 metres tall. We recognize beside it our native grey heron, a few pairs of which are now once again breeding in our immediate neighbourhood (in the Kühkopf Nature Reserve). Many heron species have specially developed decorative feathers on the shoulder or at the back of the head, and for these distinctive features they are hunted all over the world. The herons also include the bittern and the little bittern, inhabitants of reedy areas beside our rivers and ponds. In Africa, the whale-headed stork or shoebili lives in the large lakes and swamps where the Nile has its source. The Swedish author Bengt Berg devoted a book to it under its Arabic name "Abu Markub" and so made the bird world-famous. The special shape of the beak helps in the capture of special fishes sharing this bird's environment.
11 Beside the familiar white stork we see the much rarer black stork, an inhabitant of secluded forest regions. The spoonbill makes a striking impression because of its flattened beak. It occasionally visits West Germany from its breeding grounds in Holland. "A philosopher among birds" is a fanciful description often applied to the bald-headed marabou for it will remain still for hours on end with its head sunk between its shoulders, giving the appearance of a thinker lost to the world. In fact, in its home regions in Africa, it is highly active, an excellent flier and an important scavenger of carrion and rubbish. Since it can take up a very large quantity of food at one time, the succeeding period of digestion is often correspondingly long.

The flamingos have strangely bent beaks which, like a duck's beak, are constructed for the purpose of filtering food-rich muds. But the relation-

White stork

ship of upper to lower jaw is switched around in this case: when a flamingo plunges its beak into mud it is the flat upper jaw that is positioned underneath. Brilliant colours are seen in the scarlet ibis from the warm regions of the Americas. It is much sought after for feathers to be used in decoration.

Waterfowl

12 Among the waterfowl on display we find geese, swans, ducks and mergansers. Mute swans are familiar to us in gardens and parks. They still occur wild or as escapees in many parts of eastern Germany but they are more common as tame birds in parks. The somewhat smaller whooper swan, with its black and yellow beak is a common winter visitor from the far north. The striking black-necked swan from South America and the black swan from Australia are common in ponds of zoological gardens. The Cape barrow geese have a strange appearance with their short high beak, their powerful legs and the almost entire lack of webbing between their toes. They do much more running than swimming and, emergencies apart, scarcely ever enter water. They share their "quarrelsome natures" with the Egyptian goose. The common shelduck is a frequent inhabitant of our coasts, where its colourful appearance causes it to stand out among the more sombre gulls, sandpipers and plovers. It usually nests in holes in the ground or in some other concealed site. It shares this habit with the cinnamon brown ruddy shelduck, whose home is in Central Asia, but whose wanderings not even rarely bring it to the Mediterranean countries and to us. In captivity it is very tame. The Indian bar-headed goose, the magpie goose with its deeply divided webs, the Magellan goose and the small brightly coloured rust-red, black and white red-breasted goose are often on display in zoos.

Black-necked swan

The whistling geese, a group which comprises only 8 species, includes the red-billed whistling

107

Teal

Garganey

10

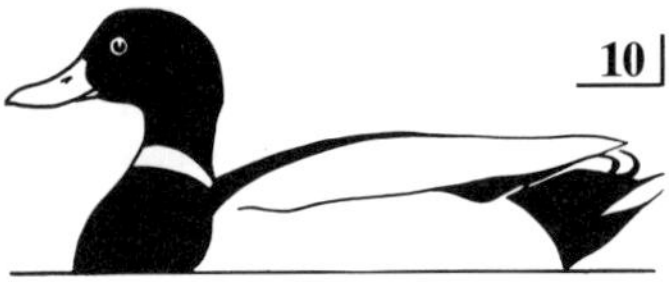

Mallard

Shelduck

duck and the white-faced whistling duck. Since they quite commonly alight on trees they are sometimes called "tree ducks". They usually breed on the ground, or else in tree-stumps or holes in trees. The smallest of all the duck-like birds is the African pygmy goose. Finally we note familiar forms, the bean goose, the white-fronted goose and the brent goose. The grey-lag goose which breeds in northern and eastern Germany among other places is the wild antecedent of most of the strains of domestic geese. The spurred or spur-winged geese from Africa have a characteristic carpal spur on the wing.

Our native surface-feeding (dabbling) ducks sit much higher in the water than the diving ducks and do not immerse their tails. They are represented by the northern mallard, the teal and the garganey, the latter two of which are our smallest species. They stand beside the long-tailed pintail, the European wigeon and the gadwall. The northern shoveler has a deformed beak and strikingly colourful feathers. The northern mallard is regarded as the root-stock of most domesticated ducks. The steamer duck gets its name from its strange mode of locomotion. It is flightless and when it uses its short wings in swimming it creates a great deal of noise and splashing. During the mating season the moscovy duck emits a strong musky odour from one of its glands.

In the Australian musk duck a fold of skin is drawn between the branches of the lower jaw, which gives the bird's head a bizarre appearance. The brightly coloured Carolina wood duck sits beside a nest in a hole in a tree out of which the young have just emerged. Its counterpart in eastern Asia, even more colourful, is the mandarin duck. These two species, often reared here, bring a decorative effect even to small streams and ponds.

If we turn now to the diving ducks, we should mention only the black scoter and the velvet scoter, both of which can be found in the North Sea region, the European pochard, the greater scaup, the common golden-eye, and the tufted duck, all of which usually appear in masses on our lakes during winter, and the eider duck from the far north whose astonishingly dense, soft belly feathers are used as eiderdown. We see here one common eider sitting on its eggs and the nest is surrounded with these soft brownish feathers. The mergansers are duck-like birds with slender, sharply pointed, hooked beaks which they use to catch fish. They are often to be seen together with ducks on our rivers and lakes during cold winters.

10

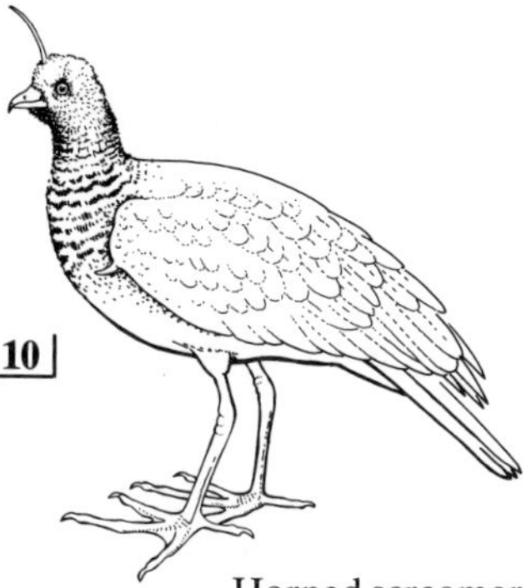

Horned screamer

Screamers

This section concludes with two representatives of the screamers which are regarded as relatives of the geese and ducks because of their way of life and their possession of certain characteristic details of body form. All of them have powerful spurs at the wrist of each wing. One species has a hornlike projection on its forehead which is, however, nothing other than a specially developed feather.

Diurnal Birds of Prey

The largest of the birds of prey, the raptors, is the condor, whose wingspan can reach 3 metres. Its weak beak, bald head and only lightly hooked claws all show it to be a vulture which feeds only on dead animals and cannot cope with living ones. These characteristics apply to most of the other vultures: the long-necked griffon vulture and the dark brown cinereus vulture from the Mediterranean region, the Indian black vulture which haunts the "Towers of Silence" where they feed on the bodies of the dead, and the black vulture or urubu which scavenges the warm regions of the Americas. The bearded vulture (also

13

Griffon vulture

106

Secretary bird

Peregrin falcon

Honey buzzard

called the lammergeier) which is no longer to be found in the Alpine region but is still present in mountainous regions of southern Europe, Africa and Asia, will take living prey too: young chamois, mountain goats, sheep which it drives over precipices, and to destruction, with powerful blows from its wings. The harpy is the most powerful bird of prey in the New World. It preys chiefly on the monkeys and sloths of the South American jungles. The long-legged secretary bird from the plains of Africa makes, by comparison, almost a peaceful impression although some of its activities are highly dangerous: it preys on snakes, even the most poisonous ones. Its name refers to the bunches of feathers developed at the back of its head, which lend it the appearance of a scribe with an old-fashioned quill pen set behind his ear. The powerful form of the African
15 bataleur is named for its aerobatics. Our native birds of prey are the buzzards, kites, the hawks, the sparrowhawks, falcons and harriers. Next to the golden eagle, the bird which appears on the German seal, we see the bald eagle, the symbol of the United States, and the African sea eagle with its splendid contrasts of colour. We are impressed by the slender form of the peregrine falcon and the hobbies which are among the fastest fliers and can catch even the fleet doves and swallows. The nearly white gyrfalcons from Greenland were at one time the treasured hunting-birds of kings. The pure white goshwak from Australia looks almost like a smaller counterpart. To the miniature forms among raptors belong the pygmy falcons found in Africa and Malaya, which are scarcely bigger than a sparrow.

Fowl-like Birds

16 Turning to the fowl-like birds, we are struck first of all by the sight of the argus pheasant, which in

Argus pheasant (*Argusianus argus*) ▶

mating display raises its wings high so that its form is concealed as if under a feathery cloak. Hundreds of "eyes" come into view artfully arranged in series on the broad feathers, and two almost 2 metres long tail feathers tower above all else. The rare Rheinart's crested argus, a brilliant item in any collection, drags a tail almost equally long. Beside it, the Congo peacock almost fades into insignificance, although it is an even greater rarity. First discovered in 1936 in the jungles of the Congo and, as a relict form, now in danger of extinction, it is rarely represented in museums and zoological gardens. We know the rock partridges from the Alps and the Mediterranean countries. The obviously larger snow partridge is an inhabitant of the high mountain country of Central Asia. Among our own native field birds, the partridges and quails, the latter is now, unfortunately, only rarely seen. Partridges, on the other hand, are more frequently seen, but the increasingly technological character of modern agricultural practice, the spread of herbicides and progressive removal of hedges and bushes all operate to this bird's disadvantage. Examples of the colourful pheasants deserving special note are the Swinhoe's pheasant, the Reeve's, the Lady Amherst's and Bulwer's wattled pheasants. This last one looks like a white disc when, in mating display as shown, it brings its tail feathers over its back. The beak and face are then fully concealed behind blue flaps of skin.

Further fowl-like birds are: the Javanese green peafowl with its wonderful long tail, the wild turkey from the United States and ocellated turkey from Central America with its coppery, shimmering green feathers. Just as improbable is the feathering of the Himalayan monal pheasant, with all the colours of the rainbow: it gives sheens of metallic red, green, violet, its back is snow

14

Satyr tragopan

Satyr tragopan (*Tragopan temmincki*) ►

white and the end of its tail is a deep chestnut brown. Nearby, the satyr tragopan presents in display a grotesque appearance with its raised "horns" (in fact flaps of skin) and the broad, inflated blue throat-lappet. Mountain qualis and francolines are to be seen next to the jungle fowl, the ancestor of our domestic hens. The true ancestor is taken to be the red jungle fowl of India and the Malayan archipelago. The jungle fowl of Java and Ceylon, and the Sonnerat's jungle fowl, are regarded as separate species. What an array of different races have been bred over the centuries from this colourful wild animal, which still lives today, withdrawn in jungles of India and the remote islands of Southeast Asia.

From the guinea fowls of the African deserts and steppes we cast a glance at the tiny buttonquails which provide a link to the cranes. Remarkably, the female is the more colourful in this case. Further, it is the larger of the two, it even engages in courtship displays and the male has entire responsibility for brooding on the eggs and rearing the young. This exceptional state of affairs is found in one further case among birds: the phalaropes, small swamp birds of the far north. Very odd breeding behaviour is found in the brush turkeys, which lay their eggs into a pile of decaying leaves and other plant matter. The heat generated by the rotting vegetation then brings the eggs to hatch as if in an incubator. Such nesting heaps can be up to 1 metre high and have 3 metre diameters. Their internal temperatures and humidity are constantly regulated by the male keeping watch, the young free themselves from their dark prison and emerge fully feathered. The curcassows and guans are the South American equivalent of the fowl of the Old World.

The tinamous also live in South America. In spite of their external resemblance to fowl-like birds the tinamous do not belong to that group. Several

Prairie chicken

Prairie chicken (*Tympanuchus cupido*)

characteristics suggest that they are an isolated group of birds which can be regarded as the most primitive family of today birds. The ostriches are probably their nearest living relatives. A noteworthy feature is the shiny, porcellanous appearance of their eggs.

The grouse native to Europe include the hazel **18**
hen, the capercaillie shown here with its hen, the black grouse and the ptarmigan exhibited in different kinds of plumage. There follows a group of prairie chickens from the "Wild West" of North America. The male, in display, with two orange-like inflated sacs projecting from either side of the neck has a striking appearance.

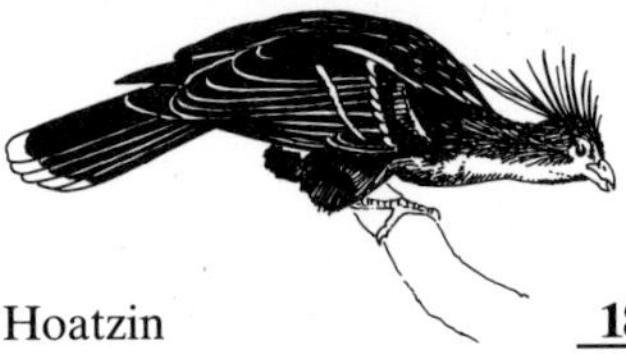

Hoatzin 18

Crane 17

Great bustard

The hoatzin is notable for the fact that at young stages its wings still have functioning claws by means of which the sparsely feathered young bird, not yet capable of flight, can climb around the tree in which its nest lies with extraordinary agility.

Cranes

We now meet the cranes, long-legged dwellers in lonely swamp and moorland areas, one species of which is counted among our native birds. In autumn it is regularly seen flying in wedge formation with loud trumpeting calls, as it makes its way through the skies to Africa. The largest form is the Sarus crane of Southeast Asia, an almost pure white species is the Manchurian crane of eastern Asia. The crown crane, whose head decoration stands up like horny spikes, comes from Africa. A characteristic of some cranes is a kind of dance-display, performance of which is, however, not confined to the mating season. A deviant form is the nearly flightless kagu of New Caledonia, where it is close to extinction. Specimens are very occasionally to be seen in zoos.

In the great bustard, some of the white underfeathers are everted during the last stages of mating rituals to cover almost all of the outer feathers. The bird then looks like a moving bundle of white feathers as it swaggers back and forward. Bustards were at one time widespread in our region. They are still to be found in Saxony, the Magdeburg area and the Brandenburg plain, where they occupy the wide swampy areas in the main. Their weight ranges up to 16 kg and they are, with swans, our heaviest birds. An array of related forms inhabit the steppes and deserts of Africa and there are a few in Asia and Australia, too. Some of them have distinctive decorative feathers on their heads and necks. Strange inhabitants of the jungles of tropic America are the limpkins and trumpeters. They provide a link to the rails and coots, to which group the well-known coot belongs. It is one of the commonest birds in rivers, pools and lakes, even in winter time. The other species in this group are so shy and withdrawn that only the initiated have any hope of observing them. The sun bittern of the tropical Americas displays its beautifully decorated feathers during mating, those on its wings are particularly impressive.

Shorebirds

Gulls and terns are, with only a few exceptions, inhabitants of the seashore and they are spread throughout the world. We find the black-headed gull and the black tern in inland areas during summer, but there are also other species which occasionally migrate from the coast into the hinterland or explore streams and lakes. The terns, perhaps the most elegant fliers of all, have a noteworthy ability to swoop vertically down onto the fish on which they feed.

The guillemots and auks inhabit the rocky cliffs and islands of high northern latitudes. At first glance they appear to resemble penguins, but they are in fact good fliers. The common puffin is a small form of auk. It has a very high (but narrow) coloured, parrot-like beak. Like the guillemots and razorbills it lives in the far north. The guillemots are content to lay their single egg on narrow ledges of rock, but the puffin digs deep nesting holes. On some islands off northern Russia, Norway or Iceland the populations of guillemots and auks number hundreds of thousands. They still breed on a cliff on Helgoland.

The ruff lives in wet meadowland, in north Germany especially, and engages in strange competitive games almost like the jousting tournaments of mediaeval times. Their feathers display a variety of colours and the broad ruffled collar serves as a shield. The jacanas belong in a related group. They have very long toes and are able to run on floating plants. One further remarkable

17

Tern

19

Puffin

Ruff

Oyster-catcher

bird is the Egyptian plover, which pecks out food particles from the teeth exposed in the open mouth of a resting crocodile.

We recognize among the snipes our own woodcock, which hunters call "the bird with the long face". The much smaller jack snipe is known for the bleating mating sounds it makes with its outer tail feathers during flight. The lapwing, with its distinctive crest of feathers, is a herald of spring. Many forms of shore birds swarm around visitors to our northern coast. The sandpipers lurch back and forth, with incredibly rapid tripping movements of their legs as the waves travel up and down the shore. The black and white oyster-catcher, its long red beak, is frequently seen along the North Sea coast. We note the plovers, among them the dotterel, the subject of a book by Bengt Berg.

21 **Doves**

Among the doves we are struck by the incredible play of delicate pastel shades in the fruit pigeons, whose feathers are chiefly green, yellow and wine-red, but black and white are also present. The light grey bleeding heart pigeon has on its white breast a blood red spot which looks like a wound. The masked dove or Namaqua-Dove is easily bred and kept in captivity. Equally distinctive are the blue-headed quail dove and the shiny Nicobar pigeon shown standing beside the rare tooth-billed pigeon from Samoa. The North American passenger pigeon, now extinct, was at one time present in thousands in the woods between the Hudson and the Mississippi, but man's rage to destroy put an end to them, just as it almost did to the bison herds of the prairies in the wild west. Among our native species of doves we find an example of a recent arrival, the collared turtle dove, a species deriving from India. For centuries their home was in Asia Minor and the Balkan countries; but a few decades ago, for reasons that are still not clear, they began to

Victoria crowned pigeon

spread westward and northward. They have settled in Germany, indeed throughout central Europe and are now established in Denmark, England and Sweden as well. The Victoria crowned pigeon is the largest of all in this group. The sandgrouse are almost exclusively ground-dwellers. Their colour, structure of plumage and excellent powers of flight show them to be notably well adapted to life in the deserts and steppe-lands of Africa and Asia. In their bodily structure and external form there are some dove-like traits, but their mode of life suggests more of a resemblance to that of the fowls, while many other characters show a close affinity to plovers and waders.

Parrots

The parrots are a well defined group although there may be great differences in size, colour and shape. The short-tailed are termed the parrots, the long-tailed are called the parakeets. This is only a very rough identification. An erectile crest is the most obvious feature of the cockatoos. Very impressive is the palm cockatoo, which has the largest beak of the parrots and whose white cheek spots can redden when it is excited. The kakapo from New Zealand is nocturnal. The kea from the same region is said to attack and feed on sheep occasionally, although this is not part of its usual feeding habits. The ground parrots remain on the ground almost all of the time and they nest there. The lories, magnificently colourful, roll out their slender brush-like tongues to lick sap and nectar from plants. The red-sided eclectus parrot stands out because of the different colourings of the two sexes: the male green, the female blue and red. The grey parrot or Jako and the predominantly green Amazons are well known and loved as talking pets. Small parakeet species such as the cockatiel and the budgerigar are even more popular. Budgerigars are bred by the million. The small African lovebirds, inseparably

20

Macaw

Palm cockatoo

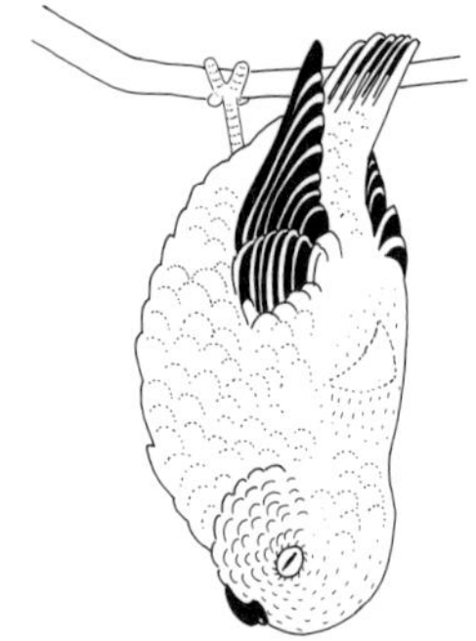

Blue-crowned parrot

Cuckoo

attached to one another, are also bred in captivity. So too are the colourful broad-tailed parakeets such as the eastern rosella or the crimson rosella. One extinct parakeet is on display here, the North American Carolina parakeet, which became extinct around the turn of this century. We can contrast with the large macaws from tropical America the dwarf among parrots, the tiny pygmy parrot from the jungles of New Guinea. Mention should also be made of the blue-crowned hanging parrot of the group of bat-like parrots, which in rest often hangs head downward – understand that the specimen on display here has not fallen from its branch!

22 **Cuckoos, Coucals and Touracos**

Among the cuckoos are our native species (here a young bird is being fed by reed warblers, its foster parents) and in addition the brilliant green emerald cuckoo, the ground-dwelling pheasant-coucal of Australia and the lizard cuckoo from the island of Jamaica. Standing beside them is the touraco, whose primaries tend to lose their colour during bathing because the reddish copper dye (turacin) can be dissolved away. Another member of this group is the brilliant deep blue Ross's touraco from the rain forests of Africa.

25 **Hummingbirds**

"Flying jewels" is a name given to the hummingbirds. They have also been called "The flower-birds of America". The colours of these tiny, swiftly flying birds are radiant. They include the smallest of all birds. Some of them are no bigger than a bumble-bee and weigh only 1.6–1.8 grams. The smallest indigenous European bird, the goldcrest, has a weight around 5–6 grams. The larger forms of hummingbird approach the size of a swallow. About 320 species are known. Most of them are found in tropical America, although during summer some forms will migrate far to the north or south, to Alaska or Patagonia, and they can be encountered in mountainous regions at heights approaching 5,000 metres. The small species breathe four to six times per second and their daily intake of food can be estimated at almost twice their own weight. They alone among birds can suspend themselves in the air like insects, holding a position, in front of a flower, say, for several seconds. Thus, they achieve by maintaining their wing beat at a rate of 20–25 (large species) to 50–55 (small species) per second – compare 13 beats per second in the sparrow and 80–90 in the case of the hornet. Their name refers to the humming sound which results from this rapid wing action. In large areas of the regions they inhabit they assume the function of the pollinating insects in that when stretching our their long tongues to suck nectar from a flower they take up pollen and transfer it to other plants. They vie with the birds of paradise in the variety too, in the form of their beaks, representing adaptations to the particular forms of flowers which individual species of humming birds prefer.

Nightjars and Swifts 25

There is only one native species of the nightjar. Folk wisdom confers on this bird the name "goat sucker". Its feathering is bark-like, and thus camouflaged, it is not easily detected either on the ground or when it lays itself along the branch of a tree. Some African and South American species are distinguished by elongate feathers in the tail and in the wings. The great frogmouth is perhaps the largest representative of this group. The oilbird is especially interesting in that it is an inhabitant of dark caves and uses echo-location, like a bat, to orientate itself. Among the swifts we note the native swift. The small swiftlets use quickly solidifying saliva to bind the so-called edible nests they build on rock faces.

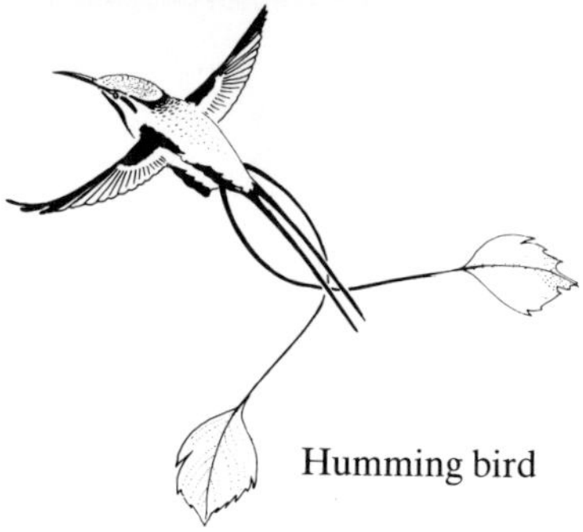

Humming bird

Nightjar

Swiftlet

Coraciiforms

25 Nearby, there are several species of the bright-coloured bee-eaters, among them also the European one.

26 The best known of the hoopoes is our indigenous form, the "bird with the feathered crest". Under threat, the young birds produce secretions from their uropygial gland. The distinctly unpleasant smell surrounding the nest hole arises for this reason and is not the result of fouling of the nest. We may admire among the kingfishers the striking turquoise green native form, which has now become relatively rare, and also the tiny three-toed kingfisher and, providing a contrast of size, the Australian laughing kookaburra whose ringing call has earned it its name. Among the rollers one notes the motmot, in which the shafts of the two long middle tail feathers are bare and hang down like the pendulum of a grandfather clock. The bird itself produces this effect in cleaning – it strips away the vanes of these feathers. The hornbill bears on its beak a large excrescence which is light and filled with airin almost all species. Possibly it has a role in mating display. During the period of incubation of the eggs the female of some hornbill species moults, withdraws, aided by the male into its cavity in a tree and is cared for there by its attentive mate. The huge ground hornbill is remarkable for its long eyelashes.

Hoopoe

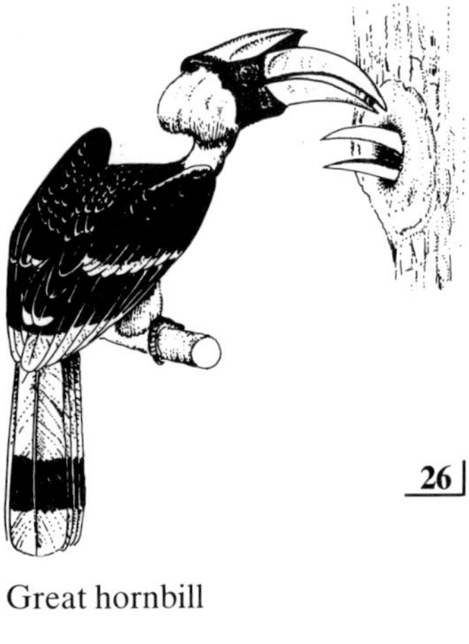

Great hornbill

Owls

26 The collection of owls includes all of the native species. We also see forms which occasionally stray into our region from northern and eastern Europe such as the lapland owl, the snowy owl and the hawk owl. Many of these regularly wintered in East Prussia and at times even bred there. The eagle owl, once present in all of the central European forests has sadly become very rare and is now to be found at only a few localities in our country. Incomprehensibly, owls are still regarded as bringers of bad luck in many

places and this superstition often costs them their lives, in spite of the fact that there is no better exterminator of pests. The burrowing owl of America lives in the underground lairs dug by animals such as prairie dogs, anteaters and armadillos and indeed may on occasion share their dwelling with them. The fish owl of Southeast Asia shows a peculiar form of adaptation to its habits in feeding: its toes have small pins of horn which ensure a more secure grasp of its slippery prey.

27

Eagle owl

Woodpeckers and Related Forms

The group of woodpeckers and related forms includes the frequently colourful barbets of tropical regions in both the Old and the New World, the brightly iridescent jacamars, the rather dull coloured puffbirds and the American toucans with their huge, paper-thin beaks. Familiar figures are the native black, green and grey woodpeckers – they are the "carpenters of the woods" – as well as the great middle and lesser spotted woodpeckers. Among foreign forms, the tiny white-barred piculet of South America deserves special note. Beside it we see the Euro-African wryneck, so named because of the moves it makes in its attempts to intimidate and drive off any enemies who may have approached the hole in the tree where it makes its nest. The magnificent shimmering green quetzal, the divine bird of the ancient Aztecs, is the most impressive member of the trogons.
The mousebirds are a small African family of six species. Their phylogenetic relationship are still enigmatic.

27

Black woodpecker

Perching Birds or Passerines

The pittas, strikingly coloured, unusually short-tailed ground-dwelling birds from tropic regions in the Old World, introduce the most extensive order of birds, that of the perching birds. All others among the birds on display in this hall are

105

still members of this same group, which includes more than half of all known birds. If it seems to the layman to be astonishing that one related group can display such a variety of forms, it should be born in mind that the many details of anatomy which they have in common serve as evidence of close kinship, even if the largest species weighs as much as 1700 grams and the smallest only 4.

A small group of birds, the manakins from the tropic regions of the New World, have only recently received close study. The crowns of their heads show a variety of colours, white, red, yellow, blue, black or opalescent, and this colouration is important in mating displays.

Next to them are the tyrant flycatchers, some of them with very long tails and feathery crests, representing a large American family of birds which is in some instances reminiscent of the flycatchers, tits and larks. The tyranny indicated in the name is a reference to their aggressive attitude toward other birds. The cotingas include the bare-throated bellbird and the Costa Rica bellbird, which produce extraordinarily loud, resonant calls with an almost metallic ring in them. They have curious excrescences on their throats and beaks. Another member of this group is the splendidly colourful cock-of-the-rock, whose upper beak is covered with feathers.

The ovenbird family includes a wide range of species. The one representative on display is shown together with its large clayey nest.

Lyrebird

28

The lyrebird looks more like a fowl than a perching bird. It has an impressively large, spread-out tail, and is an astonishing gifted imitator of any sound, whether issued by man, animal or instrument. We generally think of crows and ravens as monotonously black fellows of different sizes. But it is not widely known that this group includes also the jays and magpies and that there are, among these such strikingly colourful forms as the blue and green jays, the blue magpie and

our indigenous jays. Even among the ravens there are some multicoloured species.
From the wattlebirds, a group with 3 forest-dwelling species of New Zealand, the kokako and the extinct huia are on display. The huia is unique among birds, both sexes differ sharply in the shape of the bill: comparatively short and strong in the male but long and curved in the female. They are believed to have searched for food in the low forest level in pairs only.
The next group is the shrikes. It includes our native species such as the red-backed shrike, the woodchat shrike, the great grey shrike, also the beautiful red Burchell's gonolek of Africa. The splendid birds of paradise are closely related to the ravens. The magnificence of the feathering can only be appreciated when it is deployed in its mating array. There are ornamental feathers on the head, neck, back, wings and tail which supply a lavish variety of form and colour. Nature has revealed a wealth of inspiration in these partly bizarre, partly wonderfully delicate creatures and one response has come in the form of man who has sought to seize these forms and colours for his own ornament. Bird of paradise plumes and even whole skins were so much in demand for fashions and decorations that some species were close to extinction at the time when exports from New Guinea, the home of these birds, were at last strictly controlled. We mention here only the greater and lesser birds of paradise with their fluttering yellow decorative feathers, the striking blue bird of paradise, the long-tailed species of the genus *Astrapia,* the red and white king bird of paradise and the king of Saxony bird of paradise with feathers of especially bizarre form on its head, breast and tail.
29 We see beside the weaverbirds, starlings and troupials some of the small Australian, Indian and African finches which are collectively called waxbills and are among the most prized birds raised by devoted breeders. Three weaverbird

Bird of paradise

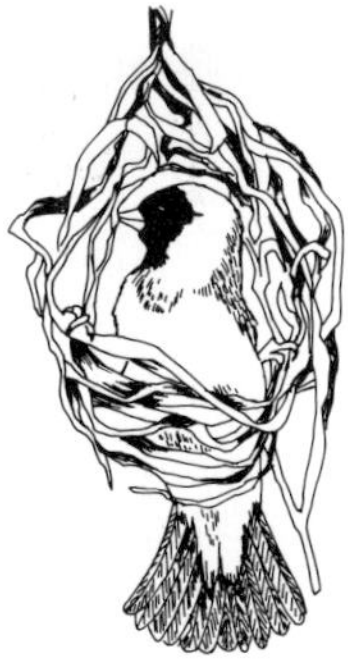

Weaver bird

105

species are on display, with their cunningly plaited nests, large numbers of which usually hang down, closely adjacent to one another, on trees and bushes. Weaverbirds, however, build a number of nests for play or for sleep in addition to the nests in which they lay their eggs, or they may produce only half-finished nests, content simply to have exercised their urge to weave. Some troupials build similar nests. This is a group of birds, confined to the New World, which in spite of their external resemblance to starlings are not related to them. Among the starlings there are some splendid green forms with a metallic sheen, the glossy starlings of Africa, and there is also the Indian hill-mynah, an excellent mimic of human voices often kept as a pet.

Indian hill-mynah

The oriole family comprises about 34 brightly coloured species found in Africa, Eurasia and Australia. The only European representative of the group is the yellow and black oriole.

31 A survey of the various related groups within the perching birds would not be complete without a reference to the tiny, shimmeringly colourful sunbirds of the tropic areas, the boldly trustful (and therefore well-known) titmice, the treecreepers, nuthatches, and goldcrests of European gardens and woods. The plain coloured pipits are also shown, and the long-tailed wagtails, the wrens, the dippers and a winter guest from the north, the waxwing. Many of them are on display with their nests or with a representation of their natural setting, for example the long-tailed tit and the bearded tit.

Wren

30 In the display of European and American finches many of the species are shown with their nests or with some indication of their preferred environments. The group includes the buntings, cardinals and tanagers as well as the true finches. The tanagers' feathering is especially colourful. They, like the cardinals, live in North and South America. A representative species (from Madagascar) of the family of the widespread small green

white-eyes is shown. This group takes its name from the white surround of its eyes. Adjoining them we find representatives of Australian honey-eaters, in which the fanned-out tips of their tongues are to be regarded as an adaptation useful in the process of feeding on fruit sap, honey and insects.

We close this account of the rich variety of the bird life which lies before our eyes at home and
abroad with a look at some small songbirds. The 33
small green tree warblers are frequent, if easily overlooked, summer visitors to our gardens and deciduous forests, the reed-warblers are tireless singers among rushes and reeds. They have skilfully constructed hanging nests which can, however, also be found in cornfields and meadows. Redstarts and robins are friendly companions even in the smallest gardens. Beside the plain-looking queen among singers, the nightingale, there is the best of the foreign songbirds, the Indian shama. We also note the compactly built native thrushes, one of which, the blackbird, is among our commonest birds.

The song of the larks announces spring to us, just 32
as do the returning flights of swallows. Our display brings out clearly the different points of appearance and habit which distinguish the several kinds of swallow found in our region. Sand martins can be seen at their nesting holes in a cliff, house martins have an almost completely enclosed nest under the protecting roof. The flycatchers, too, are among our commonest birds but are relatively inconspicuous and therefore less well-known. This is again true of the warblers which are to be found in gardens and among bushy trees everywhere. It is easier to recognize them from their widely varying and often highly melodious song than from their rather drab, grey-brown plumage. Beside them the Indian gold-fronted leafbird, grey-green in colour, and the Peking nightingale deserve mention. They are inhabitants of the tropics, splendid singers and

House martin

immediately trusting, apparently almost tame in their behaviour. They make enjoyable pets and are widely to be found in birdcages and in aviaries.

Extinct birds and mammals

Among the birds and mammals on display on the 1st Floor, there are a few species that were exterminated by man and thus do not exist any more: Dodo, Great Auk, Moa, American passenger Pigeon, huia, Carolina parakeet, quagga, Tasmanian wolf and Steller's sea cow (see Room 8). Many of them are especially described and illustrated in the loose leaf folder No. 1: "Ausgerottete Vögel und Säugetiere" in German, which can be obtained at the admission desk or ordered from the office of the Senckenberg Nature Research Society. These descriptions also explain the reasons held responsible for the exterminations. Thus, this folder is recommended to the visitor who wants to inform himself more comprehensively.

Room 108, Palaeobotany and Botany (Exhibition Planned)

Fossil plants

Fossil remains of plants are abundant in terrestrial deposits, less so in sediments laid down in the sea. Plant remains in quantity may, if sealed from contact with the air, change into peat, brown coal or coal itself and so can be of considerable economic significance. It is possible, too, that algal plant material was involved in the process which gave rise to petroleum.

The body of a plant is clearly composed of several parts. These tend to part company when the plant deceases and it is only rarely that we find fossilized examples of complete specimens. In most cases we encounter individual leaves, or seeds or fragments of the stem. The mode of preservation can vary widely. In the most common case only an impression of the plant is preserved. But in other cases the plant remain turned to coal and is now called a compression. It still contains information of the tissues of leaves. Wood and seeds are often found to have suffered very little alteration and can be studied using the same methods which would be applied to living examples of plant tissues. Genuine petrification is seen in the case of plant material whose tissues have been silicified or calcified by precipitation of minerals. "Fossil forests" have been discovered at many places in the world, for example near Chemnitz, the source of the fossil tree-trunks mounted outside the Museum. The finest details of structure can be preserved in fossils of this kind, and may be studied in thin sections under the microscope.

The oldest records of life come from blue-green algae and organisms resembling bacteria. They have been preserved for us in reef-like stromatolites and in small remnants of organic material enclosed in sedimentary rocks more

Tree-like lycopod *Lepidodendron aculeatum*

than 3,000,000,000 years old. These, the simplest of organisms, are made up of only slightly differentiated cells, and it seems that it was after the passage of roughly one billion years that more highly organized cells with a nucleus, comparable to the cellular components of almost all of today's plants and animals, first arose.

But a long stretch of time had to pass until at some point in the late Silurian, less than 500 million years ago, the first vascular plants began to colonize the surface of the land. Their most primitive representatives, the psilophytes, had as yet no leaves. That part of them emerging above the surface was either naked or else was provided with thorn-like outgrowths.

During the middle part of the Devonian more complex forms arose, to be regarded as forerunners of the modern pteridophytes (horse-tails, club mosses and ferns). These were dominant forms in the vegetation of Carboniferous time and grew as giant tree-like lycopod *Lepidodendron*, tree-like calamites and ferns. Associated with them were cordaites and seed ferns with fernlike leaves, which were the first seed-plants. In Mesozoic time the spore-bearing plants yielded pride of place to the gymnosperms. Of these, only the conifers still have a significant role at the present day. The cycads, by comparison, are near extinction: only a few genera and species remain. These and the *Ginkgo*, found only in eastern Asia today, are relics of families which were widespread and richly various during the Mesozoic. Two other groups, the bennettites and the caytoniales had protective housings like ovaries though clearly different in structure for their seeds.

The angiosperms, our flowering plants, appeared unannounced early in Cretaceous time and have continued to the present as the dominant group of plants. Their earliest representatives, strange in character, are not easily compared with plants living nowadays. More and more of the genera we know today emerged during the course of Tertiary time. We can infer from them that early in the Tertiary the climate at our latitude was of a warmth appropriate to the present tropic regions. Even as late as Lower Miocene, as evidence from the Frankfurt region (near Münzenberg) suggests, the cinnamon and palms were still present. Toward the end of the Tertiary, temperatures became lower. Plants of this episode were still represented in our region today, as the leaves of the "glass herbarium" from the Pliocene of Frankfurt-Niederrad clearly demonstrate. During the subsequent Ice Age in Central Europe most of the plants of the Tertiary flora died out. Pollen grains contained in peats provide us with

Reconstruction of a bennettite

Fern

an exact record of changes in the vegetation and of the return of plants to our region after the glacial episode.

Present-day Plants

Among plants living at the present day it is clear that on land, the angiosperms have established dominance. In this group the ovules are contained within a bud ovary which provides an effective protection from the world outside. After the egg-cell has been fertilized the ovule develops into seed and the ovary becomes a fruit. The seed is not simply a germ cell, but contains also a store of nutriment which provides for the growing embryo. Transport of pollen grains, which contain the male germ cells, to bring them into contact with the female germ cells sited in the ovary, is usually carried out by animals, although there are some angiosperm families which rely entirely on the wind as an agent of transport. The adaptation of flowers to suit various insects, or birds, or even field mice, is impressive. The growth of coloured displays (the flowers themselves) and the secretion of nectar are two clear indications of the partnership that binds plants and animals.

Animals may be important again in contriving wide distribution of fruits and seeds. Fruits are eaten by animals and the seeds, which are indigestible, are thereby scattered over a wider area. Burdocks become fixed to fur and hair and sticky seeds become attached to the beaks of birds. Spreading by water and by wind are also important.

Excepting only those regions which are always covered by ice, there is no area on land and no stretch of water which is not occupied by some

Wild rose

◄

Fossil *Ginkgoites huttoni* and leaves of present day *Ginkgo biloba*

specialized group of plants or another. In the sea, the dominant group is the algae, unimaginable numbers of which make up the phytoplankton. On land, the most obvious plants are the grasses, which occupy huge areas of prairie, savannah and steppe, and the trees, which in different zones produce the deciduous and the evergreen forests. The plant species living at the present day can be subdivided into several major groups. Bacteria, algae, fungi, lichens, mosses and ferns are the groups which belong among the spore-bearing plants. These stand in contrast to the gymnosperms and the angiosperms, our flowering plants. The angiosperms are divided into two groups on the basis of the number of cotyledons present in their embryos. The two groups are the monocotyledons and dicotyledons, and they can be traced back to one ancestral stock. Both have become highly specialized at the present day.
Innumerable plant specimens await discovery, especially among the spore-bearing plants. For this reason it is difficult to estimate the total number of plant species. It may be suggested that there are, today, about 250,000 species of spore-bearing plants and 350,000 species of flowering plants. Only about 2,000–3,000 species are grown for use or decoration and the number of those which are of decisive importance for feeding man and domesticated animals is as low as one hundred.

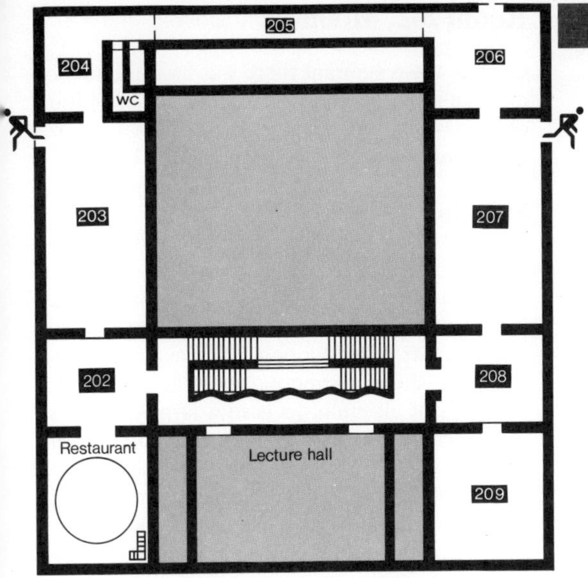

Second Floor: Fish and Evertebrates

Room 202	Mollusks (Permanent Exhibition Planned) Temporary Exhibition Agates
Room 203	Arthropods (General) and Insects (Permanent Exhibition under Construction)
Room 204	Crustaceans and Chelicerata (Permanent Exhibition under Construction)
Room 205	Dioramas (Habitat Groups)
Room 206	Marine Evertebrates (Permanent Exhibition under Construction)
Room 207	Bony Fishes and Aquarium
Room 208	Jawless and Cartilaginous Fishes
Room 209	Temporary Exhibitions

Restaurant adjacent to Room 202

Room 202, Mollusks

Outside the restaurant there was an exhibition of modern mollusks until the temporary exhibition of agates was set up. The mollusks will be returned to this room.

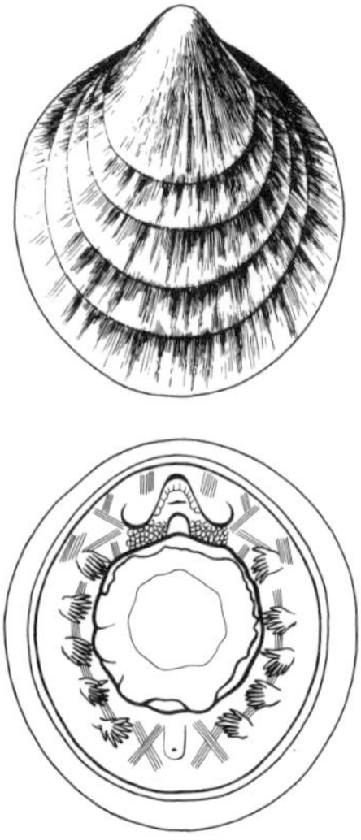
Archaic *Neopilina*

Mollusks

The first case contains on overview of the various groups of mollusks: chitons (Polyplacophora), solenogasters (Solenogastres), monoplacophores (Monoplacophora), snails (Gastropoda), cephalopods (Cephalopoda), clams (Bivalvia), and tusk shells (Scaphopoda). The anatomical structure of all mollusks is to a large extent the same, characteristically consisting of a head (mostly with a rasping tongue), a muscular foot, an intestinal sac, and a mantle cavity (mostly with gills). The mural next to the restaurant entrance shows that the mollusks form a phylogenetic unit and come from a common ancestor. This must have been a worm-like animal having a regularly metameric body with a cuticomuscular tube filled with fluid. Apart from the mollusks, annelids, which have remained worm-like, polychaetes, oligochaetes and leeches can also be traced back to this precursor.

Important steps in the evolution of the mollusks were firstly the formation of a strongly muscled foot, relocation of the body cavity to the dorsal side, and the secretion of a stiff supporting and protective shell. The modern chitons and solenogasters descend from this evolutionary stage. A further development took place with the fusing of the originally segmented shell into a unitary one such as can be seen in *Neopilina*. The most highly developed mollusk groups have various different kinds of shells: clams have shells with two valves enclosing the soft body, scaphopod shells are open tubes, while snail shells are usually a spiral coil open only at the orifice. Cuttlefish shells serve as floats for swimming, sometimes as

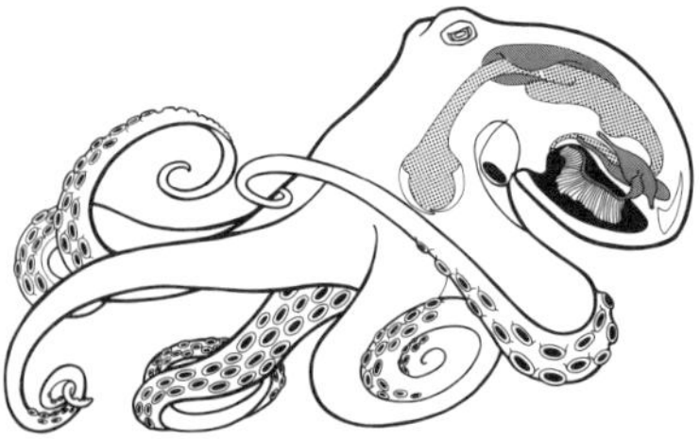
Diagram of a cephalopod

Octopus (*Octopus vulgaris*)

an external shell as in *Nautilus,* and sometimes as internal stiffening as in *Sepia*. Some mollusks have only much reduced shells and in others it has disappeared altogether (nudibranches, octopuses). The shells of snails, clams and cuttlefish

Agate gem with head of Augustus

Dendrite agate

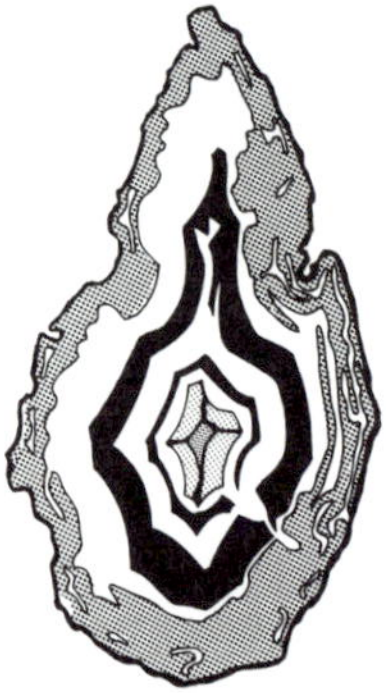

Brazilian banded agate

are often very beautifully colored and also quite resistant since they are formed of calcium carbonate. Consequently they have been used for ages either in their natural state or polished as ornaments, or for ritual and cultural purposes (case at the entrance to the restaurant). Cowries, for example, were the only currency in many regions for centuries.

From the mezzanine in the restaurant, the mural depicting the main kinds of mollusks can be seen especially well. It indicates the multiplicity of modern forms of snails, chitons, cuttlefish and clams. The large yellow diagrammatic pictures clearly show the course of the water stream which runs through the mantle cavity bringing oxygen to the gills. Clams also use this water stream for feeding by filtering out fine particles of food with their gills. Cuttlefish employ the water stream in locomotion by forcing it out of the abdominal cavity and so producing jet propulsion.

Temporary Exhibition of Agates

Agates have attracted man for thousands of years because of their unique coloration and their many varieties. Thus, the Sumerians (– 3,000 B. C.) used seals and gems cut from agates. Among ancient Egyptians, Greeks, Etruscans and Romans gems and objects of art carved from agates were very popular. Collecting of these objects in ancient Rome had reached a climax that later was never surpassed. In medieval times the church and noble courts accumulated precious agate gems and vessels in their treasure chambers. Between the time of the Renaissance and the 19th century the center of the difficult art of gem carving was in Paris.

It was not only the fine colours and manifold structures that made agates so popular, it was also, dating back to old ages, the mystic powers assigned to these precious stones that made them attractive to people of all ages.

The temporary exhibit comprises 15 cases, several maps and other picture-graphs. Case no. 1 exhibits the agate within its mineralogical frame as a representative of the chalcedons within the quartz group. Then follows a collection of typical agates of the most important locations in the world.

One priority group (3 cases) includes agates from Idar-Oberstein, the long-time agate center in Germany. Several cases are devoted to individual types of agate like pipe agates, layered agates, landscape agates, and dendrite agates. Several cases exhibit agates from Brazil, presently the important agate producer; and agate geodes provide and excellent review of coloration and structures of the Brazilian type.

One case is devoted to agates from Paraiba (N.-Brazil, discovered only in 1973). This type of agate represents casts of voids between large crystals and is thus a pseudomorph. Luminosity and dispersion of colours as well as the fineness of structures is shown well in many discs by their illumination from below. All agates on exhibit are natural specimens, they are not artificially coloured.

A small booklet dealing with this exhibit of agates is available at the admission desk or by mail order (Kleine Senckenberg-Reihe, no. 12, DM 10,–).

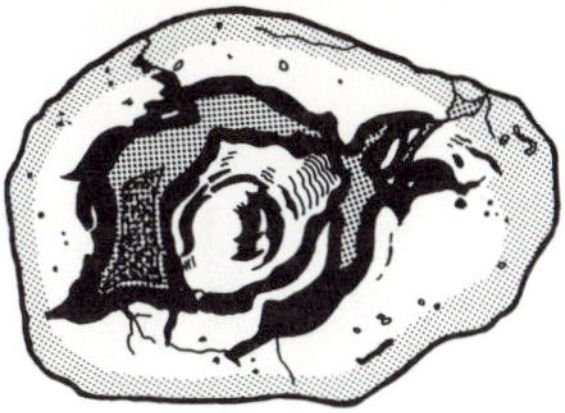

Thunderegg

Layered agate (nodule)

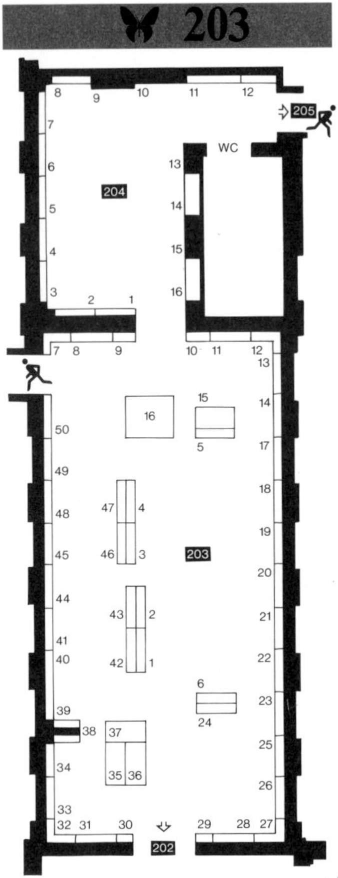

Rooms 203 and 204, Arthropod Gallery (Permanent Exhibition under Construction)

Insects, crustaceans, arachnids and myriapods together have over a million species thus forming the most extensive animal phylum, the arthropods. They are almost omnipresent in the most manifold biotopes on land and in water right down into oceanic deeps.

Room 203, Arthropods (General) and Insects (Permanent Exhibition under Construction)

Body Structure and Evolution of the Arthropods. The displays in the showcases standing in the middle of Room 203 give an introduction to the body structure and evolution of the arthropods. It is shown that there is a common principle of construction both for the adult with its hard external armor as well as for its soft-skinned developmental stages (e. g., the butterfly and caterpillar). This principle is completely different from what we are familiar with for vertebrates and for our own body where the bones, the skeletal parts, are on the inside surrounded by muscles and connective tissue and covered with a soft skin. In arthropods, by contrast, the outer surface is both the wall of the body and the
1 skeleton, in that it can contain hard plates in addition to softskinned parts.
The reason for this construction principle can only be understood in terms of the phylogenetic past of the arthropods. They are descendents of worms which were constructed in such a way that hard parts could occur in the body wall but not
2 internally. The bodily architecture of the forerunner also led to the characteristic arrangement of the arthropod body, which is often very strikingly segmented and has many legs, which are mod-

Insect gallery (Room 203)

ified into antennae and mouthparts on the head. Another consequence of the construction princi- 3
ple of the arthropod body is that the sensory organs are completely different in structure and arrangement from those of vertebrates. Insect wings are not limbs but rather membranous out- 4
growths. A huge variety of forms has developed in the course of evolution. 5

Many of the seemingly remarkable features of the anatomy and mode of life of arthropods are the results of their bodily structure. For instance, since the hard external wall cannot expand, they 6
have to molt in order to grow. As they develop

203

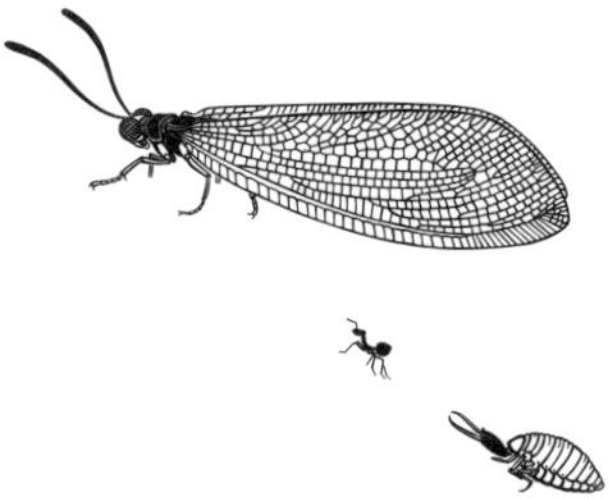

Ant lion fly: imago and "ant lion", its larval stage (below)

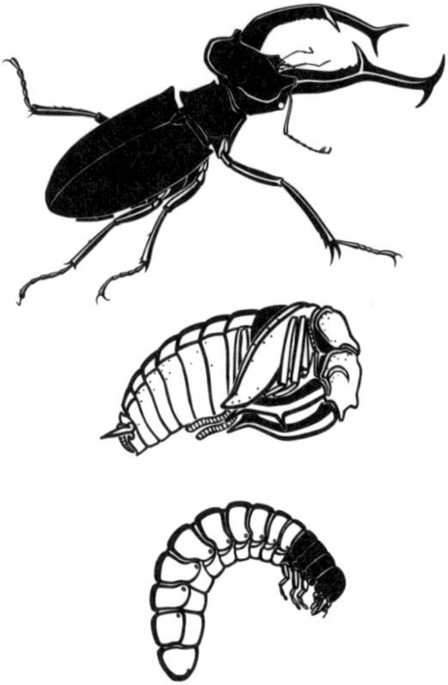

Development of the stag beetle (from bottom): larva, pupa and imago

many of them pass through a larval stage which bears no resemblance to the adult. The various groups develop in many different ways, some of which are summarized in the general overview. There are roughly two types of insect development. In hemimetabolous development the metamorphosis is incomplete, i. e., there is no pupal stage. The larvae molt several times as they grow, resembling the adult stage (imago) more closely each time until they finally attain it completely. This is supposed to be the original course of insect development. Holometabolism, or complete metamorphosis, involves a pupal stage, during which the transformation from the larva to the imago takes place. The larvae are mostly very different from the adults (e. g., caterpillar – butterfly). The details of both hemimetabolism and holometabolism can vary considerably due to ecological adaptation. Some peculiarities will be pointed out in the displays of individual insect groups.

Insects

The insects comprise by far the largest group of animals in terms of species numbers as they amount to 75% of all known species of the entire animal kingdom. Approximately 1 million insect species have been described so far. By contrast, the total number of all vertebrate species is only barely 50,000. The oldest insect remains, preserved only as imprints, come from Lower Devonian beds almost 400 million years old. Since that time, insects have developed a huge number of forms not even approximated by any other group of animals. Thousands upon thousands of species have long since become extinct, including real giants with wingspans of more than a ½ metre.
The larger part of the room is taken up by a systematic depiction of the insects. Nevertheless, due to the immense number of insect forms, a great deal can only be presented in outline. Graphics and texts in and between the showcases

are intended above all to acquaint the visitor more closely with the insects' way of life. The graphic frieze along the top of the wall indicates the largest and most important groups and provides quick orientation from a distance.
To get an overall picture, it is best to start at case 7 next to the exit to the staircase and then proceed clockwise to the right.

Apterygota

7–8

The first insect groups you encounter in this way used to be classified together as Apterygota until it came to be accepted that this did not correspond the real phylogenetic relationships. "Apterygota" means "wingless" indicating that these insects neither have nor ever did have wings. Without doubt it is correct to regard winglessness as a more primitive feature than being winged. The far-reaching agreements between the young and adult stages as well as the fact that the adults can still molt are also held to be primitive characteristics. An essential distinguishing criterion is provided by the mouthparts. Some groups have these in a sort of pocket drawn into the head capsule. These groups are called Entognatha or Entotropha, while the rest, which have the mouthparts outside the head, are termed Ectognatha or Ectotropha. The former are considered to have the more specialized mouth, whereas the structure of their antennae is held to be primitive since all sections of the antennae, except the last, are provided with muscles. The latter group, on the other hand, has more primitively arranged mouthparts but more highly developed antennae with muscles on the first section only. The majority of "primitive insects" belong to the Entognatha group while the rest, including for instance the silverfish and all winged insects (Pterygota) are Ectognatha. As this example teaches, the degree of resemblance is not always a measure of phylogenetic relatedness.

Springtail

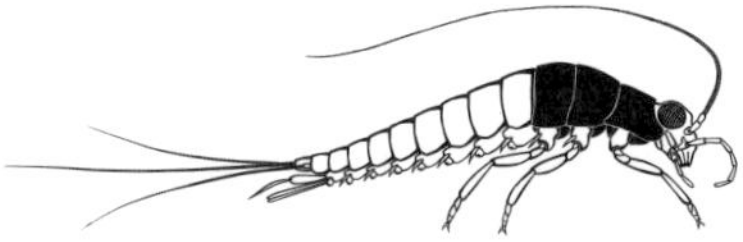
Bristletail

203

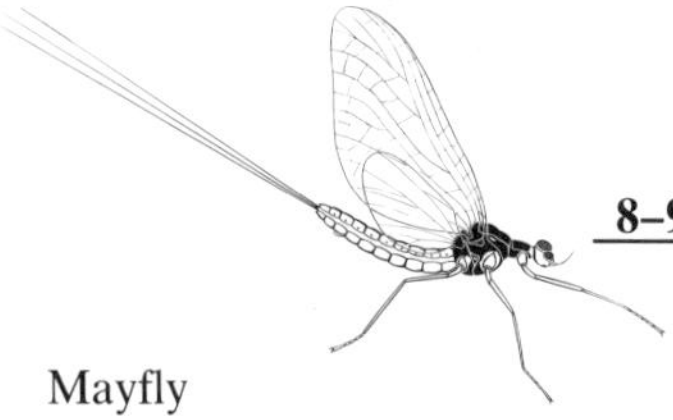

Mayfly

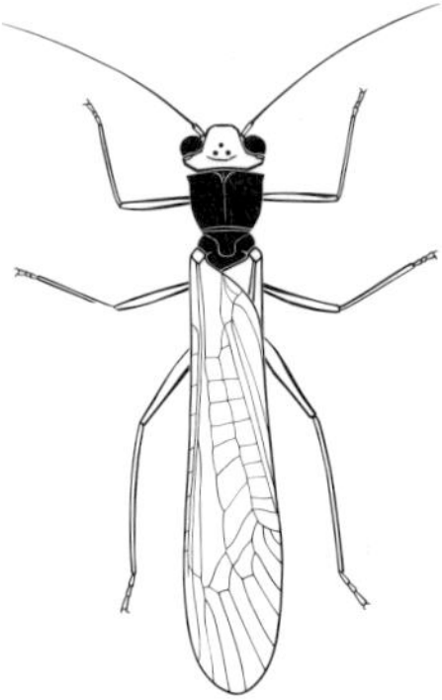

Stonefly

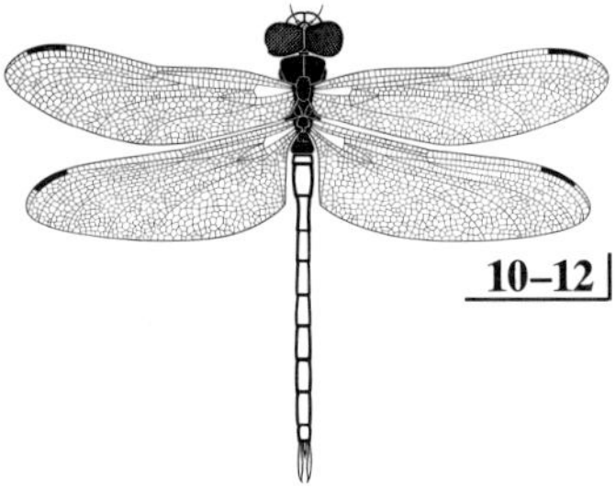

Dragonfly

"Primitive insects" rarely attract our attention and yet there are enormous numbers of them. Many of them influence the character of the soil decisively and so are of great practical significance for us.

8–9

Mayflies (Ephemeroptera)
Mayflies are delicate, slender and small. They have two or three hairlike filaments on their posteriors which assist in their slow, oscillating flight. They can very often be seen swarming near water on warm summer evenings. Their larvae live in the beds of rivers or lakes and breathe through tracheal gills which can be pinnate in some species. While the imagoes often live for only a very short time, the larval development can last a year or more. Mayflies are the only winged insects which still molt in the form capable of flight.

8–9

Stoneflies (Plecoptera)
Stoneflies live near rivers and creeks. They are more powerfully built than mayflies and brownish to yellowish-green in color. They are found settled on tree trunks and rocks with their wings folded flat on their bodies. The males mostly have smaller wings than the females, and in some cases there are short-winged variants in addition to normal-winged ones. Stone fly larvae live in water rich in oxygen; they breathe through tracheal gills and are carnivorous with a preference for eating mayfly larvae.

10–12

Dragonflies and Damselflies (Odonata)
Odonates are all exceptionally good fliers; both the adults and larvae are carnivorous. Large eyes, easily turnable heads, powerful, biting mouthparts, and forelegs well-adapted to holding prey make it easy for them to catch and eat other insects. The larva possesses a special organ which can be rapidly shot out to catch its prey of small crustacea, insect larvae or young fish. This is the

labium modified at the tip into two pincers for grasping prey.

Embiids (Embiodea)
The embiids are capable of spinning; the spinning glands are located in the first segment of the feet and open into hollow bristles on the "sole" of the first and second segments of the feet of the forelegs. They use these to build tubular webs in which they spend most of their time. The females are always wingless, the males often so. Some species have developed brood care and simple forms of social behavior. They live in mild or warm climates and do not occur in Central Europe.

Notoptera
These remarkable insects, which seem quite primitive in many respects, were only discovered quite recently in high mountain areas of North America and Japan. The few known species seem to be the relics of a once larger, more widely distributed group. Their lifespan is strikingly long. They usually do not hatch until one year after the eggs have been laid, then the larvae take about five years, in which they molt eight times, to mature into the imago. The adults then live another 1–2 years, though the males mostly live shorter since they are eaten by females after mating. Some of their features indicate relatedness with the orthopterans (i. e., grasshoppers etc).

Earwigs (Dermaptera)
Earwigs have a very movable abdomen with a pair of forceps-like cerci, which are used in various ways, for instance, to hold large pieces of food, to unfold the complicated hind wings from beneath the short rigid forewings covering them, and also in defense. Their pinch, however, is too weak to be painful to humans. Some species are unable to fly. Most of them live in groups so they

203

13

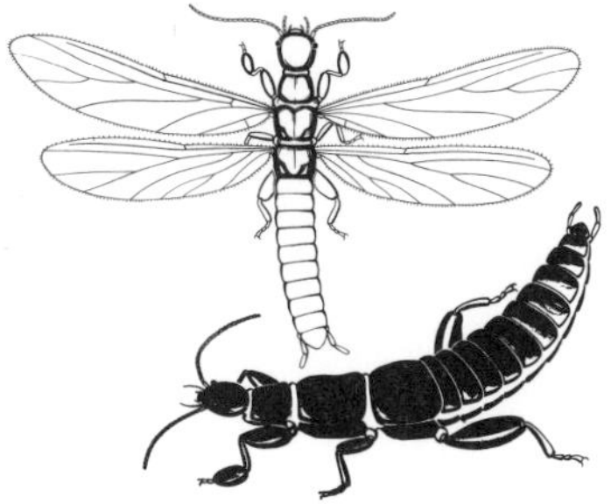

Embiids or web-spinners

13

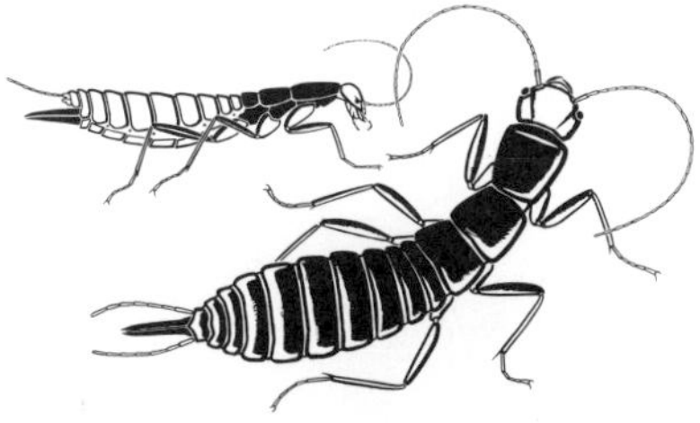

Notopters

13–14

Earwigs

are often found in large numbers in their hiding places. They have well-developed forms of caring for their brood and young. Some species are viviparous (i. e., they produce living young instead of eggs). Plants are their main source of food.

13–14 **Mantids** (Manteodea)
The forelegs of mantids are modified into arms for catching their prey which consists mainly of insects. Even the youngest larvae are carnivorous. Most species inhabit warmer climates, although one species, the praying mantis, comes into climatically favored areas of Austria and South Germany. Some tropical species mimic the shape and color of flowers which attract insects. The egg clutch is enveloped in an ootheca made of secretions from accessory glands of the sex organs which harden on contact with air.

13–14 **Cockroaches** (Blattaria)
There were already representative of this order in the Carboniferous period. Cockroaches can run swiftly but either do not readily fly or are entirely incapable of flying. If present, the forewings are narrower and more solid than the posterior wings which are folded together over the back when not in use. Most species shun the light and can crawl into very narrow cracks thanks to their very flat bodies. Despite their preference for the dark, they have well developed eyes. They can perceive sound waves using the "hearing hairs" on their abdominal appendages, and react extremely sensitvely to vibrations. A widespread characteristic is symbiosis with bacteria or flagellates which helps make food which is difficult to digest (e. g., wood) accessible to them. In dangerous situations they can discard whole legs or parts of them (autotomy). The larvae grow these again after the next molt. The eggs are deposited in an ootheca; some species are viviparous.

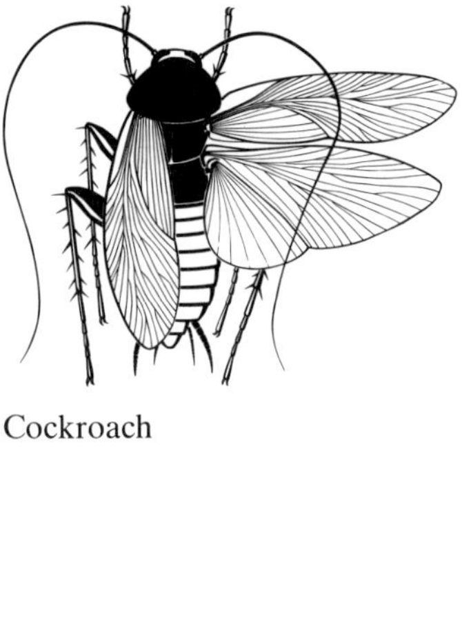
Cockroach

15 **Termites** (Isoptera)

These insects, which are evidently closely related to the cockroaches, occur only in species forming colonies ranging in organization from the simple to the highly complicated. In contrast to the colony-forming Hymenoptera, both sexes are always present in the population. In addition to the individuals capable of breeding, usually a “royal pair”, there may be different castes for different tasks (e. g., workers, soldiers). Since termites are hemimetabolous, the larvae also take part in the work in the colony. The queen is chiefly occupied in egg-laying in which her performance is quite incredible. The largest number of eggs laid on one day by one female has been observed in the species *Odontotermes obesus,* it was 86,400! No wonder that the abdomens of the queens swell up to many times their original size. Many species erect voluminous nest with complicated interiors.
16 The cast of a nest built by the Australian compass termite is on view in the display. All termites are herbivorous and, like some cockroaches, their digestive processes are supported by symbiotic flagellates. Some species cultivate fungi for use as food.

17–18 **Stick and Leaf Insects** (Phasmida)

This group includes a particularly large number of bizarrely shaped species. Some look like twigs, others like leaves. These are evidently defensive camouflages. Such adaptations of the body shape to structures normally occurring frequently in an animal's biotope are termed mimesis or mimicry. Stick and leaf insects enhance this effect by their slow movements and ability to become rigid. Although they have no saltatorial legs, stick and leaf insects are reminiscent of grasshoppers, with which they are most likely related, in other respects such as mouthparts, ovipositor, wing venation etc. The ability to fly is either very limited or absent. Some species, such as the walking sticks often kept as pets, reproduce parthenogenetically.

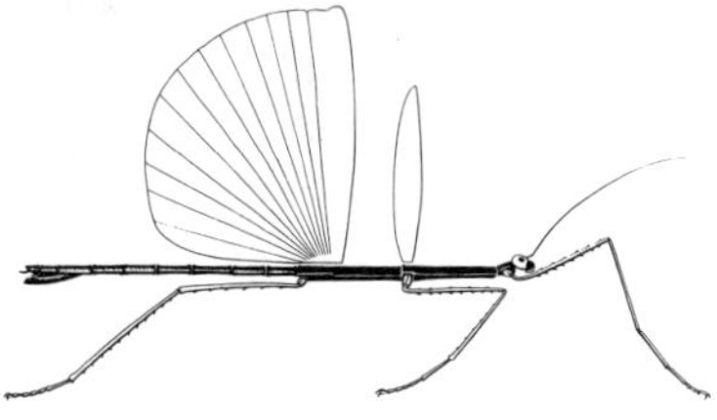
Walking stick

Ensifera

17–18

The Latin name of this group, Ensifera (literally, "sword bearer") refers to the ovipositor of the female, which forms a striking, straight or saber-like curved bodily appendage in many species. Further distinguishing features are the long antennae, the stridulating organ at the base of the anterior wings and the tympanal organ on the tibia of the forelegs. One or the other of these structures may be retrogressive in any particular species. Not least due to their "song", various species of ensiferans such as the large green grasshopper from the bush cricket family, the field cricket, and the house cricket have become popularly known. The song is produced mostly by the males, less often by the females also, and mainly serves to locate the sexes. Each species has its own "song" which the members recognize innately. Species with reduced wings usually retain wing stumps with the stridulatory organ. The eggs are buried in the ground or in a plant by means of the ovipositor. With a few exceptions, the larvae of Central European species do not hatch until the following year.

Some ensiferans are entirely herbivorous, others prey on other insects, other species again are less specialized and have a wide food spectrum. An example of the latter kind is the mole cricket which mostly lives underground and is astonishingly well-adapted to digging.

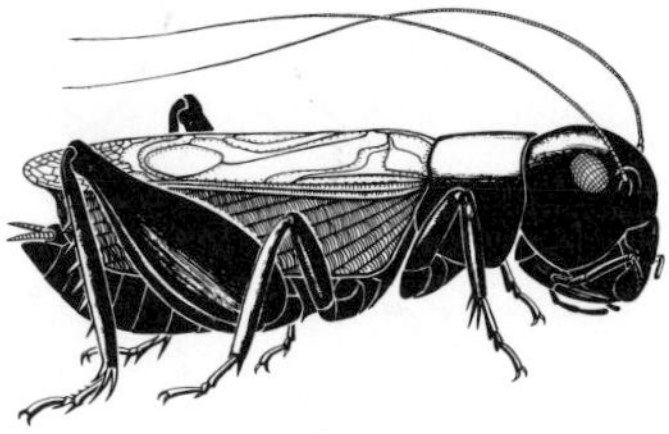

Field cricket

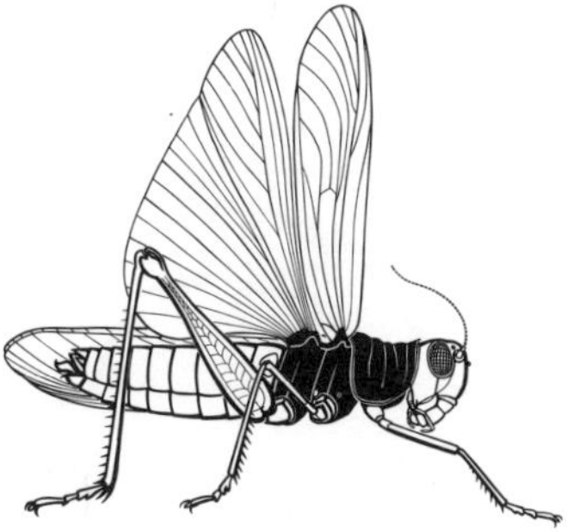

Grasshopper

Caelifera

17–18

At first glance caeliferans resemble members of the preceding group, however, their antennae are usually much shorter and taper off less sharply. If they have tympanal organs at all these are located on either side of the first abdominal segment. They have no long ovipositor. The song – many

◀

Indo-malayan mantid (*Deroplatys* sp.)

203

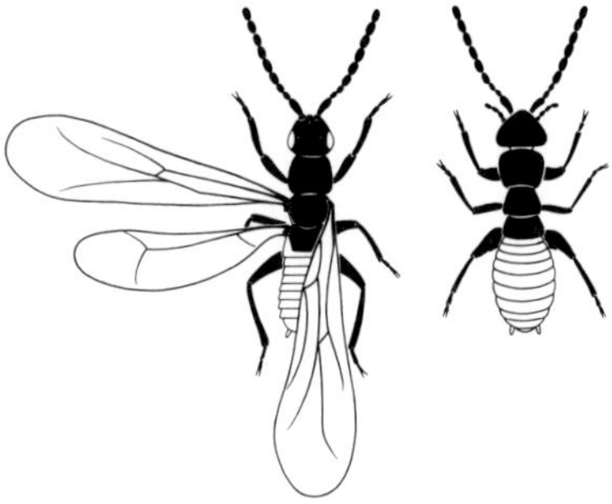

Zorapterans

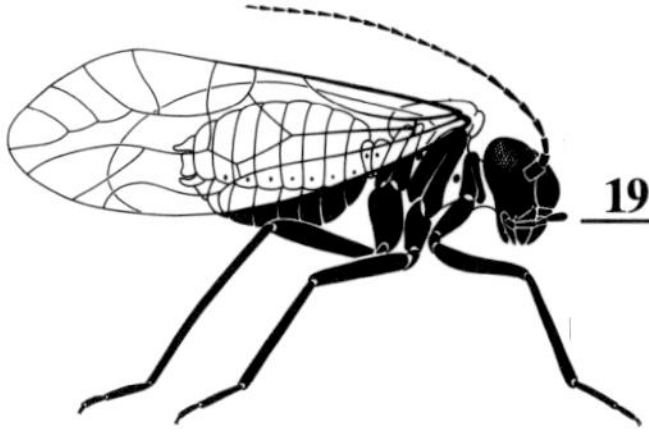

Booklouse

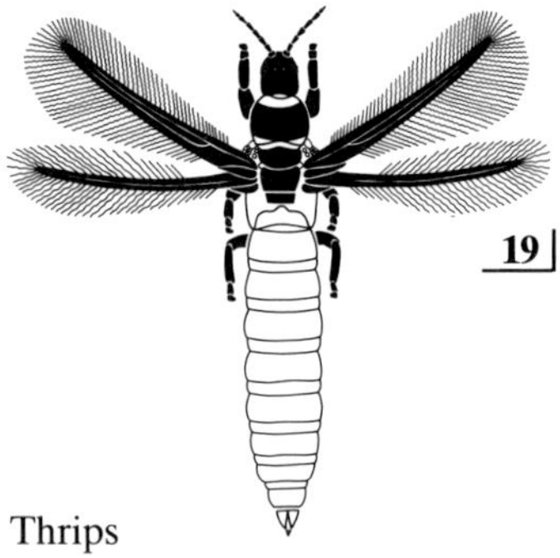

Thrips

species have their own special one – is produced by rubbing the femurs of the hindlegs over a particular area of the wing – the details of the stridulatory organ vary according to the species. In addition, they have other ways of producing sounds: rattling the wings while flying, rubbing the mandibles against each other, drumming with the hindlegs. Their posterior wings are often colored very strikingly, while the rest of their coloration usually provides excellent camouflage. Some species are even able to change their color so as to blend into their surroundings. Caeliferans are almost without exception herbivorous. Some species, such as the notorious locusts, can cause enormous damage at times of mass incidence. The species known in the vernacular as the bush cricket, which populates the meadows in summer, also belongs to the caeliferans. Most species deposit their eggs in the ground, a few in plant tissue or in balls of secreted matter glued between grass blades.

19 **Zorapterans** (Zoraptera)
So far only 24 species of this insect order are known. They are small, mostly colorless forms which are found in all tropical regions living in groups on the ground or under tree bark and probably feeding on fungi.

19 **Booklice** (Psocoptera)
These small, hemimetabolous insects live mostly on plants (tree bark, dead branches); many species also occur in houses. Their food consists chiefly of fungi, algae and lichen.

19 **Thrips** (Thysanoptera)
The thrips are small, hemimetabolous insects which mostly feed on plant saps. When they occur in large masses they can damage crops by juice sucking and also be annoying to humans (e. g., grain thrips).

Animal lice (Phthiraptera)

19

All members of this order live as ectoparasites on other animals. They are all wingless and have gripping feet which ensure them a steady hold in the hair or feathers of their hosts. Two subgroups can be distinguished: sucking lice (Anoplura) and biting lice (Mallophaga). The former have piercing mouthparts and suck blood, the latter eat the horny surface layer of feathers, hair, and skin. Sucking lice are found only on mammals; monotremes and marsupials as well as some other groups (e. g., bears and cats) have no lice. Three species can occur on humans and may transmit dangerous diseases (e. g., typhus, relapsing fever). Biting lice live mainly on birds, more rarely on mammals. Like sucking lice, they are very host-specific, i. e., certain lice species are found only on certain host species.

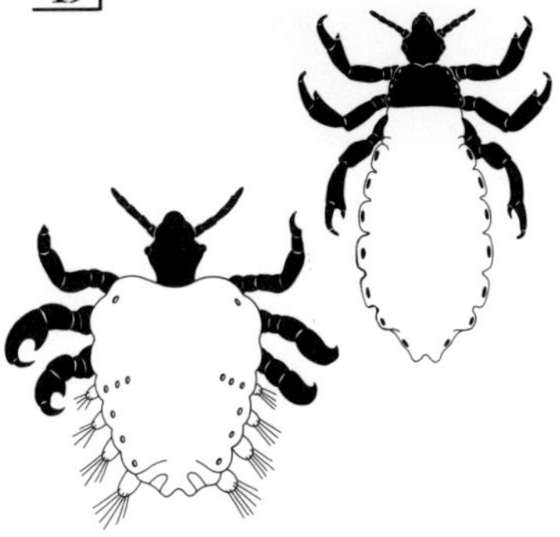

Lice on man: crab louse (left) and head louse (right)

True Bugs (Heteroptera)

19–20

Because of the well-known pest, the bedbug, most people have negative opinions about this group. Nevertheless, the Heteroptera, of which there are about 40,000 species, by no means all deserve our aversion. Some species, such as brown seedbugs, pyrrhocorid bugs, and striped bugs are even quite handsome. Since their forewings are often thickened at the base, laymen often take them for beetles. A second characteristic of true bugs, besides the wing development just mentioned, is their piercing mouthparts used by the various species to suck sap from plants or blood from animals. This piercing proboscis can also be used in defense, and a sting by some of the larger species can be quite painful even for a human. Many species have adapted in various degrees to aquatic life, some as swimmers, others as water striders, which exploit the surface tension of water to be able to run about on it. Most true bugs (at least in the larval stage; they are hemimetabolous) have abdominal odor glands which produce unpleasant smelling

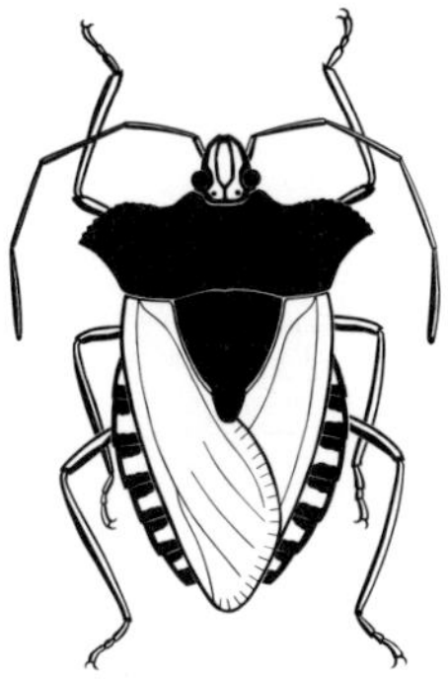

Shieldbug

203

Predatory bug

21–22

Bedbug

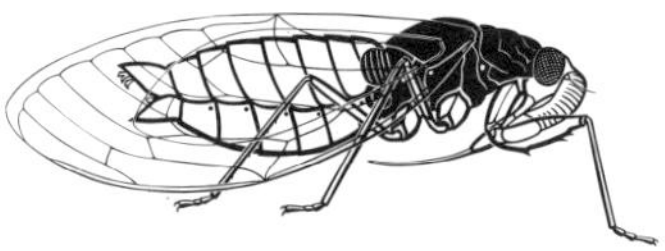
Cicada

secretions, poisonous to other insects. They are not immune to this poison themselves, but it cannot penetrate their skin, and, thanks to special protective devices at their respiratory orifices, cannot reach their tracheae, either. Various families have stridulating organs but the significance of these is not yet fully understood.

Aphids, Cicadas, etc. (Homoptera)
These plant-sucking insects used to be classified together in the order Homoptera and contrasted with the true bugs (Heteroptera). Some experts are now inclined to divide the Homoptera into at least three different orders, one of which, however, the Coleorrhyncha, includes only a few insignificant species living in South America, Australia and New Zealand. The other two groups, cicadas and aphids, are very much better known. They feed exclusively on plant saps. Their mouthparts are modified into piercing and sucking proboscises. Cicadas have short, bristle-like antennae and usually well-developed wings folded together like a gable roof over their back when at rest. Many species are able to produce sounds. The songs of the smaller forms are scarcely audible for humans, while those of the larger species are amongst the most striking sounds in nature in warmer regions. The organ for producing sounds by the vibration of a special membrane exerted by muscular action is located on the first abdominal segment, and is one of the most complicated sound-producing organs in the animal kingdom. Many cicadas can jump very well. The family of treehoppers has always attracted special attention because they have large and often weirdly shaped excrescences on the first segment of the thorax, the biological significance of which is still unclear. Many species have a quite complicated alternation of generations. Skin glands for producing secretions (wax, lac) which can form protective integuments, are widespread. Mass occurrences of some species can

lead to crop damage. Others were, or still are exploited economically (shellac production). Since the uptake of carbohydrates in plant saps is excessive, plant lice secretions often contain sugar. This honeydew is occasionally collected by other insects, above all by ants, which sometimes keep and tend regular herds of aphids. So-called fir-honey also consists of the secretions from some aphid species which live on conifers and which has been processed by honeybees.
While all the insect groups so far described are hemimetabolous, the remaining orders are holometabolous. From the viewpoint of species number, these make up by far the largest part of the insects.

203

Beetles (Coleoptera)

23–28

Beetles can be recognized at first glance by the elytra which are the sclerotized anterior wings serving as a protective cover for the posterior wings. These are of membrane and mostly much longer so that they have to be folded up under the elytra when not in use. When flying, it is the membranous wings that are active, while the elytra function at the most as airfoils more or less spread out, or even remain flat on the back as for instance in the rose-chafer.
Both pairs of wings may be reduced in size in varying degrees; the rove beetles (Staphylinidae), for instance, often have hind wings fully suitable for flying under their very short elytra. On the other hand, the posterior wings may be lacking although the elytra are fully developed (e. g., some carabids). As might be expected, this huge group of insects includes examples of the most various kinds of specialization in body form and mode of life, which cannot be described even approximately here. The mouthparts, however, are predominantly of the chewing type, though they can be employed in many different ways. Beetles can be purely herbivorous, carnivorous, or omnivorous. Most beetles can walk well, many

Common cockchafer

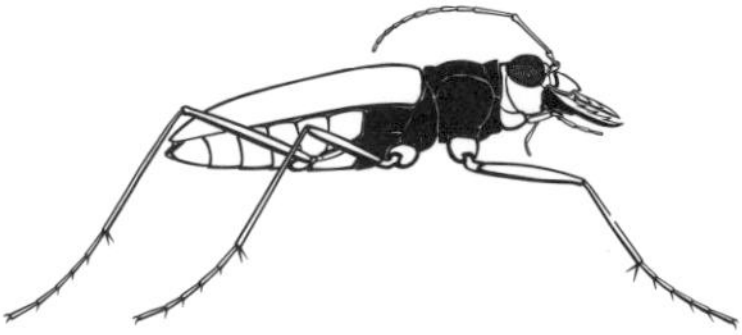

Tiger beetle

dig, while others swim using legs modified into perfect oars. Skin glands for secreting odorous and defensive substances are widespread. Some groups have developed notable forms of brood care. The larvae may be very agile with well-developed cursorial legs, but maggot-like forms are also common. There are mostly three larval stages, sometimes more, which can be very different from each other. This is termed polymetabolism or hypermetabolism. Many species are economically very important either as pests or as useful insects, a few are bred on a large scale (e. g., the larvae of various beetles sold as mealworms).

28 **Stylopids** (Strepsiptera)

This small order seems to be related to the beetles, and is classified with them by some experts. It consists of endoparasitic species whose hosts are mainly hymenopterans, though they also occur in cicadas, true bugs, flies, and even primitive insects. Only the males are capable of flying which they do with their fan-shaped posterior wings, as their forewings are reduced to appendages like small halteres. The females are always wingless and very often have no legs, antennae, or eyes, either. When this is so, the females do not leave their host, but merely protrude their anterior part when they have reached sexual maturity. The male then flies to them and mates. The larvae, born alive, are very agile in their minute first stage. They leave their mother's abdominal cavity, in which they have developed, and infest other insects. Species which live in hymenopterans have themselves brought into the host's nest where they penetrate into the bodies of the young larvae of the host and there continue their development. They do not damage vital organs in their host, but do impair its health considerably. Often the host's gonads cease to function (parasitic castration).

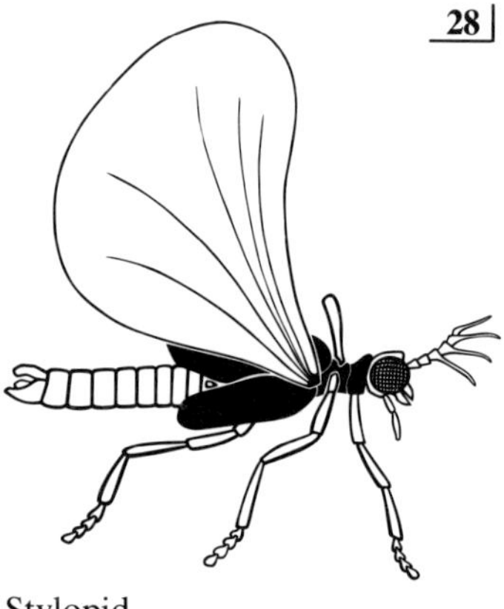

Stylopid

The Neuropteroidea comprise the following three orders of alderflies, snakeflies and lacewings (in the narrow sense).

Lacewings (Planipennia)
These insects have two pairs of narrowly veined, membranous wings with which they fly rather clumsily. Only the strikingly colored owlflies can fly somewhat better. All species are carnivorous, and their larvae are often especially efficient hunters, such as those of the common lacewing, or of the ant lion which lurk in funnels in the ground and are sometimes called doodlebugs. Mantispids are worth mention for their forelegs modified into prey-catching arms somewhat resembling those of mantids. While the larvae of some Planipennia live in water, they always pupate on land. The old larvae can spin and make a silk cocoon before pupation.

Alderflies (Megaloptera)
There are only about 100 species of alderflies. The imagoes are quite sluggish, and make little use of their long, richly veined wings. Their eggs are deposited on plants above the surface of water in such a way that the larvae can drop into the water on hatching. They have a pair of tracheal gills on the abdominal segments and prey on various small creatures. The larvae pass through ten stages in two years before going onto land to pupate in the ground.

Snakeflies (Raphidioptera)
The number of snakefly species is not much larger. They can be easily recognized by the forward elongation of the first thorax segment. They are carnivorous, their larvae invading the burrows of beetles that live in wood to eat their larvae. The individual development can last two years.

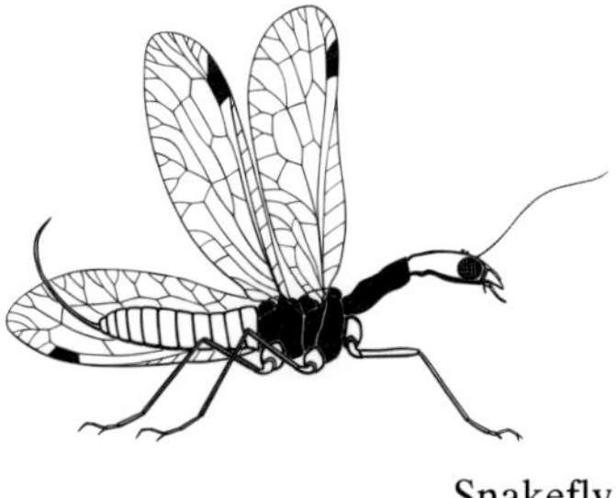

Snakefly

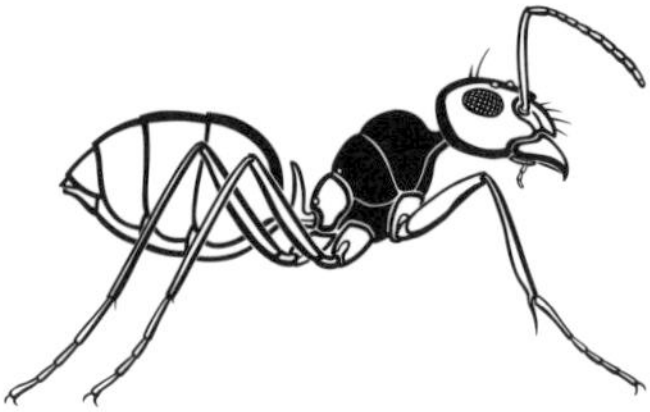

Ant

Horntail

Bumble bee

Hymenoptera

This group consists of a very large number of species which, as a rule, have two pairs of membranous, sparsely veined wings with the anterior pair larger than the posterior ones. Sometimes, as in ants, the wings are retrograde. The first abdominal segment is closely connected with the thorax, and fuses with it in the more highly evolved species, so the "wasp waist" is located between the first and the second abdominal segment. Another characteristic feature is the ovipositor, which is unique among the holometabolous insects, and rather resembles forms found in grasshoppers. It may be provided with venom glands and be modified into a very effective sting for defense. The mouthparts are adapted for licking or sucking, while the mandibles always remain powerful and form excellent biting organs. Species with the most various modes of life belong to this order. The more primitively built sawflies and horntails (no wasp waist) are predominantly herbivorous. Their larvae sometimes look very like butterfly caterpillars and live on or in various plants where they may also become noxious. The subgroup with the constricted abdomen (Apocrita) contains many more species. Among the relatives of the ichneumon flies, there are innumerable parasites on other insects, often with incredible adaptations. The smallest known insects, only 0.2 millimeters long, are to be found among the fairy flies which belong here. Their larvae develop as parasites in the eggs of other insects. Other groups such as scoliid wasps, spider wasps and sphecid wasps are carnivorous, but their imagoes also visit flowers. The forms of social organization developed are particularly diverse, some species even herd animals and others cultivate fungi. Bees have become almost exclusively feeders on flowers. Together with the termites, the hymenopterans are the only insect group with colony-forming species. Several species have developed this

Underground nest of German wasp (*Paravespula germanica*)

mode of life independently of each other. Nests 35–37
built by some social species are exhibited in the island showcase. In the summer months, the activities of a live colony of honeybees can be 38
observed. The economic significance of the hymenopterans can hardly be overestimated; no other insect group includes so many useful species.

Butterflies and Moths (Lepidoptera) 41–46

These are without doubt the most popular insects. The reason for this is certainly the colorful-

ness of their large wings as the exhibits, especially those in the island showcases, help to show. The wings are densely covered with variously shaped scales which also function as color producers both by means of pigments as well as by their structures. Moreover, the scales improve the flying characteristics of the wings. Except for primitive moths, the imagoes of most species have retrograde mouthparts. However, there is usually a proboscis formed of parts of the lower maxilla. It is often very long and is coiled up when not in use. Usually it serves for the uptake of fluids (nectar, water, and, in rare cases, blood). The imagoes of some species (e. g., goat moths) no longer take nourishment. Some species have an extraordinarily well-developed sense of smell so that they can detect a single molecule of a scent in favorable circumstances. Many species, above all the hawk or sphinx moths, are rapid, high-endurance fliers, and some even migrate over hundreds and thousands of kilometers. The wings of other species are much reduced. Their larvae are commonly known as caterpillars. Unfortunately, it is often forgotten that butterfly protection should also involve caterpillar protection. For example, anyone who wants to enjoy the peacock or small tortoiseshell butterfly also has to let nettles grow as they are the foodplant of the caterpillars of these species. Many caterpillars can spin threads which they often use during pupation (e. g., as a harness, cocoon, etc.). Silkworm cocoons have been used by humans since ancient times. Caterpillars often have camouflage coloring to make them invisible in their surroundings, whereas poisonous or ill-tasting species often have a

Swallowtail

Hawk moth

◀
South American, nymphalids (from top): *Catagramma hydaspes, C. peristera, C. sorana, C. lyca.* Each one with upper side (left) and lower side (right).

bright, distinctive warning coloration. It has been demonstrated that, when butterflies with eye-markings on their wings, suddenly open them displaying the "eyes", this causes a scare reaction in birds and so often saves the butterfly's life. For the rest, the butterflies' modes of life are extremely diverse corresponding to the gigantic number of species.

Scorpion fly

45

Mecoptera

These insects are reminiscent of flies on the one hand, and Planipennia on the other, but the proboscis-like elongation of their heads makes a good dinstinguishing feature. The larvae mostly resemble caterpillars. They feed on dead insects and small animals, but sometimes also prey on live insects or eat plant parts. One species, the scorpion fly, so called because of the peculiar way the male holds its abdomen, is partially specialized in robbing spider-webs of insects caught in them.

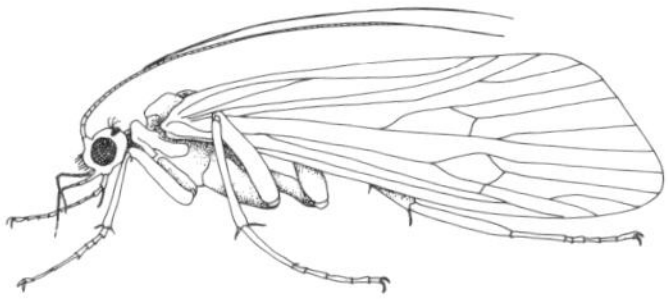

Caddisfly

45

Caddisflies (Trichoptera)

These insects are certainly very closely related to the Lepidoptera. The larvae of many species build a protective case in which they live. As a rule the larvae live in water, often attached to stones or the like. Some species go through their developmental stages in the litter on forest floors. The materials used to build the body case are very diverse (small pebbles, wood shavings, pieces of leaf, small snail shells, etc.). These are all bound together with threads spun by different parts of the body depending on the species. Some species construct thread nets which trap microorganisms that are then eaten by the larvae. The imagoes usually have two pairs of wings, more or less thickly covered in hair. When at rest these are folded on the back either flat or peaked like a gable roof. The mouthparts are adapted to licking up fluid nutrients, however, many species no longer eat in the imago stage. Since caddisflies require water of a certain quality, they can be useful as bioindicators of the water quality of creeks, ponds, etc.

Two-Winged Flies (Diptera)
As has been shown, there are species with only two wings in other groups as well, but in this order it is the rule without exceptions, apart from species such as some louse flies whose wings are missing altogether. The anterior wing pair is present, while the posterior pair occurs only in the form of halteres. The distinction into gnats and flies is based on certain morphological characeristics, e. g., gnats have filament-like antennae, while those of flies are short and composed of variously shaped sections. The species number of diptera is only exceeded by the coleopterans, lepidopterans and hymenopterans, so the diversity of forms and modes of life is very large. There are herbivores, carnivores, blood-suckers and parasites in all grades of specialization and all biotopes except the sea. The highly specialized mouthparts are adapted in various ways to licking, piercing and sucking. Diptera are extremely significant for humans. They transmit many diseases (e. g., sleeping sickness, malaria), yet by no means all are noxious. Some are even decidedly useful (e. g., tachinid flies, aphid flies). Mimicry is a strikingly widespread characteristic: many dipterans can imitate the form, coloring and movements of hymenopterans with stings such as bumblebees, wasps and honeybees.

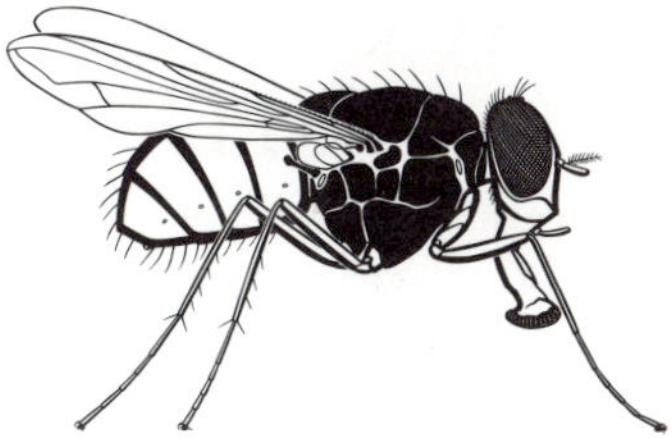

Bluebottle

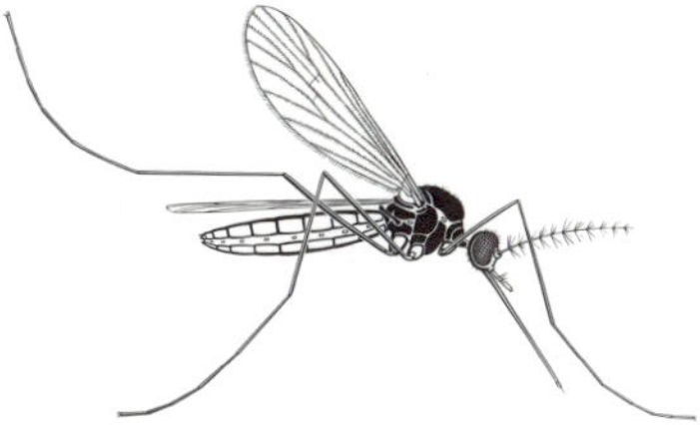

Mosquitoe

Fleas (Siphonaptera)
In the imago stage, all fleas live as bloodsuckers on warm-blooded animals, although they do not stay permanently on their hosts. They are always wingless and built so that they can easily slip through their host's hair or feathers. Particular species prefer particular host animals but they are seldom strictly host-specific. For instance, dog or cat fleas may also bite humans.

203

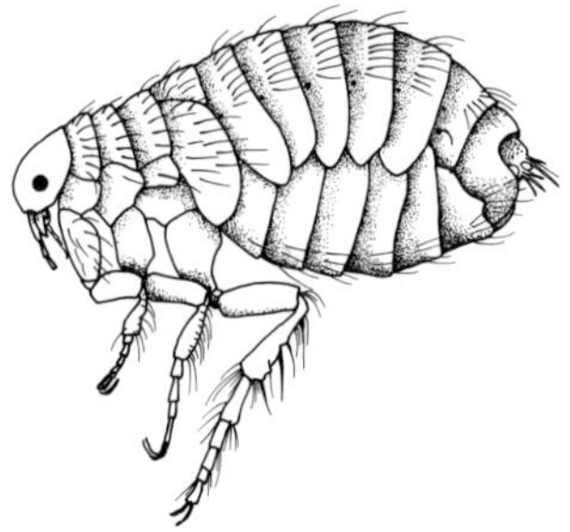
Human flea

They are known for their powerful jumping, in which they use their middle and above all back legs. The eggs are deposited so as to fall to the ground or into the host's nest, where the larvae then live on various organic substances. They pupate in a spun cocoon camouflaged with particles of materials from the surroundings. Modern methods of floor cleaning impair the larval development a great deal, so human fleas have become quite rare in large parts of Europe. Fleas can be dangerous as disease transmitters (e. g., the plague). It is not yet clear to which other groups the fleas are related, but most experts consider them as being close to the dipterans.

Room 204, Crustaceans and Chelicerata (Permanent Exhibition under Construction)

Crustaceans (Crustacea)
Crustaceans are the most successful arthropods in marine biotopes. Even though most of their roughly 35,000 species still live in the sea, many of them have managed the transition to life on land or into freshwater. All crustaceans breathe through gills and posses two pairs of antennae as organs of touch. Their gills are large-surfaced leg appendages which obtain oxygen from the water washing around them. These finely ramified breathing organs are highly susceptible to drying out and have to be kept moist, especially by land-dwellers.
Entering Room 204 from the insect room, there 16
is a showcase on the wall to the right with a display of large crustaceans. These are all decapods, immediately recognizable as crustaceans even for someone with no training in zoology. Despite all diversities, European lobster, spiny lobster, robber crab, king crab, giant Tasmanian crab, and red deep-sea crab all have a bodily structure traceable back to a common fundamental principle. The graphics between the two large 15
showcases illustrate this clearly. The "body" of the crab corresponds to the cephalothorax of the "long-tailed" crustacean forms: the abdomen (= "tail") is reduced and carried curved forward under the cephalothorax.

Giant Japanese Crab 14
The largest living crustacean is the giant Japanese crab (*Macrocheira kaempferi*). The pincers of adults of this species may span up to 3 metres when spread out sideways. They belong to the family of the spider crabs (Majidae) and inhabit a very narrowly limited area, occurring only on the Pacific coasts of the Japanese islands Honshu,

タカアシガニ
Japanische Riesenkrabbe
Macrocheira kaempferi TEMMINCK
Die japanische Riesenkrabbe ist das größte heute lebende Krebstier. Die Spannweite der seitlich abgestreckten Scheren kann bei ausgewachsenen Tieren bis 3m erreichen. Auch bei dieser Größe der Gliedmaßen beträgt die Länge des Körpers nur wenig mehr als 30cm. Der japanische Name des Tieres »Takaashi-gani« (Hochbein-Krebs) nimmt Bezug auf die große Länge der Gliedmaßen und den hochbeinigen stelzenden Gang. Die Art ist in ihrer Verbreitung auf die Küsten japanischer Inseln beschränkt und kommt dort nur auf der Pazifik-Seite der großen Inseln Honshu, Shikoku und Kyushu vor. Ihr Lebensraum sind die tiefen Sand- und Schlickgründe von 50-300m. Während sie im Sommer sich in mittleren Tiefen von ca. 200m aufhält, kommt sie im Winter auch in flacheres Wasser, wo sie sich dann in den Netzen der Fischer verfängt. Zu dieser Jahreszeit wird sie auch gegessen. Eine kommerzielle Fischerei existiert hingegen nicht, da die Riesenkrabbe doch vergleichsweise selten ist.
Das hier ausgestellte Exemplar wurde im Jahre 1902 gefangen und gelangte 1903 nach Frankfurt. Es handelt sich um ein ausgewachsenes Stück.

Shikoku, and Kyushu. In summer they live in water depths of about 200 metres, but prefer shallower waters in winter, where they then get into fishing nets. Although it is only such accidentally caught specimens which come onto the market, the species is easily endangered by this because it is so narrowly distributed. The species name of the giant Japanese crab is in honor of the German physician, Engelbert Kämpfer (1651–1716), who stayed in Japan from 1690 through 1692 and published the first illustrations of this species in his "Geschichte und Beschreibung von Japan" (History and Description of Japan). This shows how well-known the giant craw was in Japan even in those times. The specimen on display, a fully grown one, was cought in 1902, and reached Frankfurt in 1903 thanks to the efforts of the German consul in Japan at that time.

Economic Significance of Crustaceans 13

From the map of the world fishing industry, it can be seen that crustaceans are caught in nearly all seas, though the species fished depends on the geographical region. Deep sea shrimps, lobsters, and king crabs play a large part in cold seas, while spiny lobsters and prawns predominate in warmer latitudes. Shrimps make up 50% of the total yield of crustaceans so that they are the most important group economically. The graphic shows that the world yield was about 3 million tons in 1979. In view of such figures, it is not surprizing that some species are already overfished. The only way out of this dilemma is to breed crustaceans as man has done with land animals for ages. Some experiments in this direction have been encouraging, but this should not be allowed to disguise the fact that this field is

◄
Giant Japanese crab (*Macrocheira kaempferi*)

still very much in the initial stages of development.

The smaller showcases on the opposite side of the room are intended to illustrate the diversity of crustacean species and their body structures.

1 The first showcase to the left of the entrance contains the most important groups of the orders which used to be classed together as "lower crustaceans" (Entomostraca), but are now distinguished into 9 subclasses of equal rank. Special note should be taken of the primitive crustacean displayed in the upper row, which belongs to a very ancient phylogenetic group. Other specimens in this case include water fleas and cyclopids such as are well-known food for aquarium fish. The barnacles (Cirripedia) are especially interesting: their body plan is so much modified that they cannot directly be called crustaceans. The rock barnacles, goose barnacles, and parasite barnacles, which are important subgroups of barnacles, betray the fact that they belong to the crustaceans by their nauplius (i. e., larva) which is very similar to that of other crustaceans.

Water flea

2 The adjacent graphic is to illustrate that common structural principles are to be found in crustaceans despite their great diversity. This is the case, for example, in body segmentation and leg structure, even though these have quite different appearances in the various subclasses and orders.

3 **Higher Crustaceans** (Malacostraca)

A number of orders of higher crustaceans are exhibited, though no attempt at completeness is intended. Amongst the orders combined as Peracarida, the Amphipoda, and the Isopoda are quite well-known, the latter being the only crustacean group which has been very successful on land. Various species of them can be found living socially under stones and tree bark everywhere in our forests, others live in damp corners in houses (wood lice) and greenhouses. The majority, however, live in the sea, and one of them, the deepsea-dwelling *Bathynomus giganteus*, can grow up to a length of 27 centimeters, which makes it the largest known species. Of particular interest are the primitive Leptostraca and the mantis shrimps (Stomatopoda), marine carnivores with much enlarged maxillipeds modified into organs to catch their prey. The decapod order, the most successful group of crustaceans begins in the lower row. The delicately-built forms classed together as shrimps (Natania) include not only the very primitive prawns (Penaeidea), the only decapods which still have a nauplius as the larval stage, but also such highly evolved forms as the tropical freshwater shrimps, of which one large species, *Macrobrachium rosenbergii*, is on display. The next case shows the most important remaining groups of decapods, classed as Reptantia in zoological systematics. In contrast to the shrimps, this group lives on the seabed and is strongly armored. The Astacidea are characterized by well-developed pincers and a powerful, distinctly segmented abdomen. In addition to typical sea-dwellers, the best known of which is the lobster (see case 16), this group also includes the freshwater crayfish. The body shape of the Scyllaridea resembles that of the lobster, but they do not possess fully developed pincers. Tyical specimens of the spiny lobster and shovel nosed lobsters (scyllaridae families) are exhibited from this group. The second antennae pair of the latter are modified to resemble shovels, and their bodies are very much flattened. The Eryonidea are a very ancient crustacean group which used to inhabit shallow seas during the Jurassic period, but are now found only in deep seas. The Anomura occur in many forms, the best known of which are the hermit crabs which conceal their soft abdomens in empty gastropod shells. They are the ancestors of the lithodid crabs, which have evolved their crab shape independently of the true crabs – an extremely interesting case of evolutionary conver-

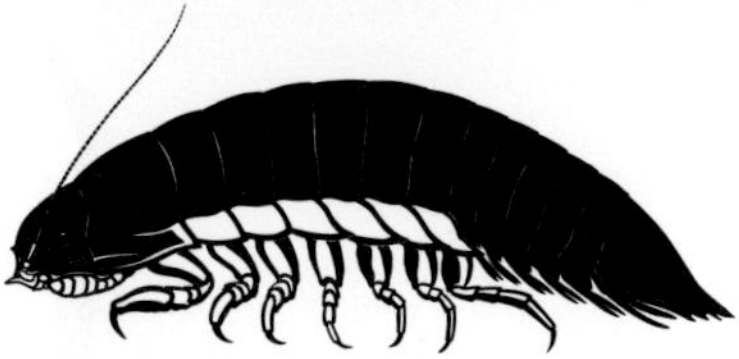

Deepsea-dwelling *Bathynomus giganteus*

4

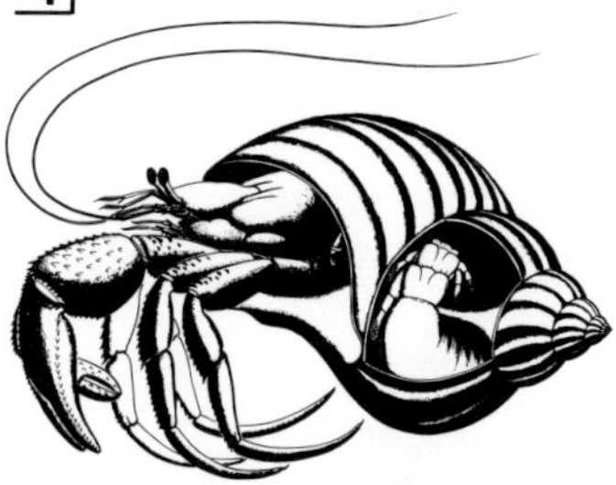

Hermit crab

Rock-dwelling crab

o
er
n
en
nst
n

cm

fi-
sind
ern. →

gence. There are about 5000 species of true crabs (Brachyura), i. e., roughly half of all the decapods. The specimens exhibited give an impression of the diversity of their forms. The relatively unspecialized rock-dwelling crabs and stone crabs (coral crab and yellow crab) are contrasted with fish hunters such as the swimming crabs.

Chelicerates (Chelicerata)
The chelicerates originated as a marine evolutionary line of arthropods. Their forerunners took up a predatory mode of life, which most of the 40,000 species have retained until the present. The first pair of legs has been modified so as to catch and tear apart their prey and has pincers or fangs (= chelicerae). This is the term which gives rise to the scientific name of the group. Only a few specimens of chelicerates are at present on display provisionally.

Horseshoe Crabs (Xiphosaura)
The horseshoe crabs (Xiphosaura) are a primitive evolutionary line of Chelicerata. They still live in the sea like their forerunners in the Cambrian period. They push the surface of the seabed aside with their broad shields and rummage through it looking for small snails and worms. The largest species, *Limulus polyphemus*, which grows up to 60 centimeters long, comes from the east coast of North America. The four Southeast Asian species remain smaller.

6

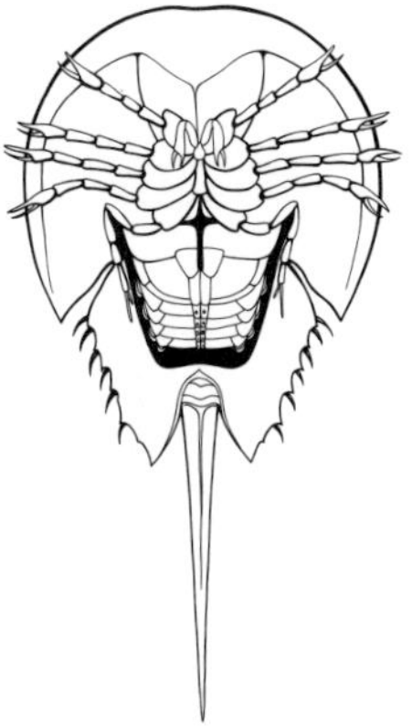

Horseshoe crab

Pantopoda
These are also sea-dwellers. The scientific name, Pantopoda (literally, "consisting entirely of legs") refers to their remarkable body structure: the legs are long, the trunk of the body is small, and the abdomen completely reduced to a stump. They have a large sucking organ like a proboscis

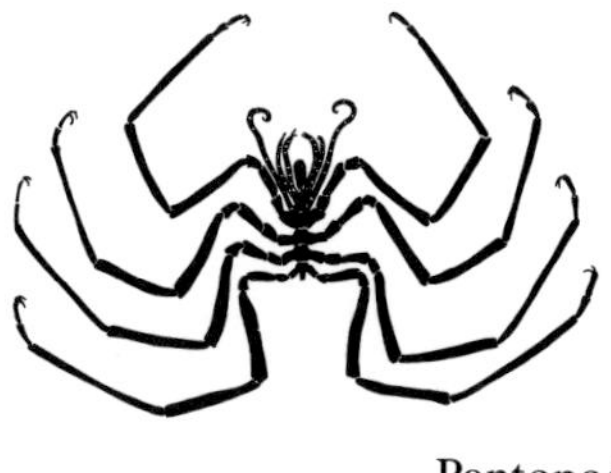

Pantopod

◄
Robber crab (*Birgus latro*)

oriented forwards. They clamber around on seaweed, coral and sponges and suck the soft parts out of sessile animals like sea anemones and moss animals. Pantopoda are one of the main evolutionary lines of the chelicerates from which they must have branched off in very early times.

Most chelicerates, however, are purely land-dwellers, which are subdivided into 9 orders of arachnids. They have six pairs of limbs: the 1st pair bears the chelicerae, the 2nd pair (pedipalps) are organs of touch, while the remaining 4 (3rd to 6th) pairs are formed into cursorial legs.

Spiders (Araneae)
Amongst the arachnids, the most important in species numbers (approx. 23,000), and consequently in the diversity of predatory life forms, are the spiders. There are scarcely imaginable numbers of them hunting mainly insects, or spinning all sorts of webs to trap them. It has been calculated, after counting up the specimens on a piece of fallow land in England, that there are about 6 million spiders per acre! All spiders are characterized by the ability to spin threads of congealing protein. The well-known orb-webs used to catch prey are examples of an especially highly developed constructional method. Many other kinds of webs, for example ceiling webs with "tripwires" to entangle insects, have less precisely arranged threads, but are by no means less successful in catching insects. Innumerable methods of using catching webs, including complicated ones, have been developed, such as webs which are not spread out until an insect approaches, or others which are dropped like a trapdoor. These more highly complicated trapping techniques, however, are used by only a relatively small proportion of spiders, most species capturing their prey on the run or by jumping. Tropical orthognathes, almost all non-poisonous for humans, and those of the genus

Orthognathe from Theraphosidae family

Nephila are amongst the largest species. Particularly bizarre and colorful forms are to be found amongst the tropical araneids.

At present, indigenous German spiders and a number of especially beautiful tropical species are on display. In addition, some orthognathes are on show, the largest of which, *Theraphosa leblondei*, is one of the most voluminous spiders of all. 7

Venomous Spiders 7

Practically all spiders are venomous, though not for humans, because the venom is normally only deadly for their prey, mostly insects, as soon as it is injected into the victim by a bite with the spider's fangs, the chelicerae. There are only a few species, 100 at an estimate, whose venom has

204

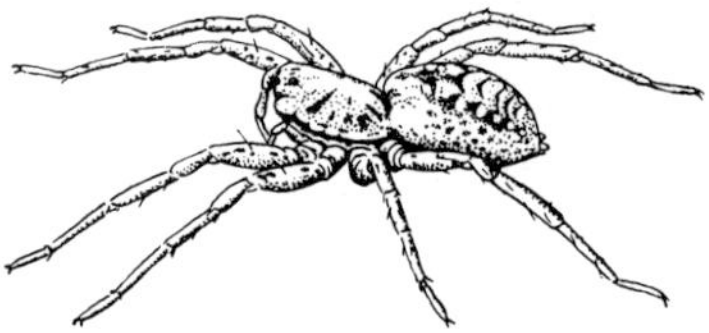

Chiracanthium punctorium

any mentionable effect on humans. Moreover, spiders are usually extremely shy of biting humans – they have to be irritated a great deal or tightly cornered to provoke them into biting. The only really venomous spider indigenous to Germany is *Chiracanthium punctorium.* Its bite initially causes a sharp pain like a wasp sting. This is followed by a purplish discoloration of the spot bitten and a swelling which lasts for days: the pain may also linger on. In addition, though not in all cases, the bite causes general indisposition, shivering fits and light fever, in one case even a circulatory collapse. This spider lives mainly in moist spots in the Odenwald-Rheinhessen region. They are particulary prone to bite in August when the females are guarding the eggsac. These spiders, however, are not common, and so few people are bitten.

It has also been reported that the bite of *Argyroneta aquatica,* which has become rare in Germany, has a certain venomous effect, likewise that of *Araneus diadematus.* Nevertheless, spider experts have been bitten dozens of times without suffering the least venomous effects: only in one case (a bite in the skin on the head) did there occur pain and a burning feeling. This shows that the consequences of a bite (as of practically all venoms) can turn out very differently. They depend on the individual sensitivity to the venom and on the patient's general constitution, and probably on that of the spider, as well. So, it can happen in rare cases that the bite of an otherwise harmless species causes a strong reaction.

The garden spider, therefore, like all other spider species indigenous to Germany (there are several hundred of them) has to be regarded, with some reservations, as harmless.

Outside of Central Europe, there are a number of quite dangerous spider species, whose bite can cause death in some cases: black widows (*Latrodectus* species) in Mediterranean lands, and Central and North America, *Phoneutria* in

Brazil, *Loxosceles* species in North and South America, *Atrax robustus* in the Sydney area of Australia, and *Harpactirella lightfooti* in South Africa. They are mostly dangerous due to neurotoxins, *Loxosceles* due to tissue destroying venom, likewise *Verrucosa audax* (Central and South America) and *Mastophora* (South America) as well as *Lycosa erythrognatha* (Brasil), whose bite causes wounds with large surfaces. Antiserums are developed to treat effects of bites.

Atrax and *Harpactirella* are the only species in this list related to the orthognathes: the rest of the numerous orthognathe species, above all the family Theraphosidae, have only a slight venomous effect on humans or are quite harmless, insofar as cases of bites are known at all.

Taken altogether, the arachnids, even if scorpions are also taken into account, are by no means the most significant venomous animals for humans as is commonly believed. Judging by the number and dangerousness of the cases, insects (bees, wasps), and snakes outnumber them by far.

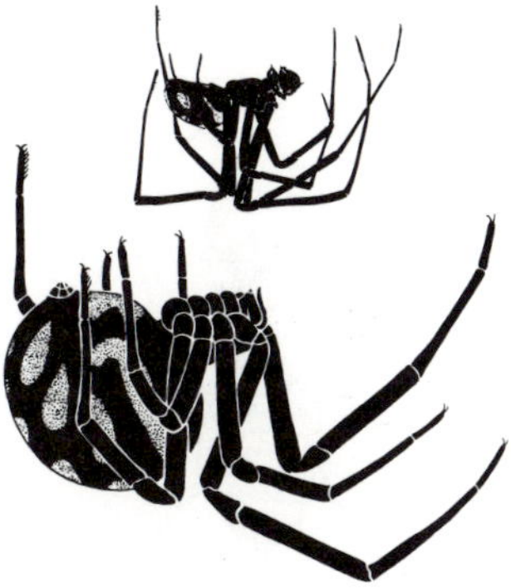

Black widow (above: male, below: female)

Scorpions (Scorpiones)

7

These are easily recognizable by their large pedipalp pincers and their slender abdominal section bearing a venomous sting. They form an integrated functional unit: the prey is held with the pincers and the body curved forward to apply the deadly sting. Only a few of the total of 650 species are dangerous to humans. These species occur in America, North and South Africa and in Arabian countries. The most dangerous species is *Androctonus australis* (North Africa) which is responsible for 80% of all cases of stings, and 95% of all deaths (treatment with a serum is necessary). Scorpions are predominantly inhabitants of the tropics and subtropics. During the day, they conceal themselves under stones, in cracks, and in holes in the ground. In desert

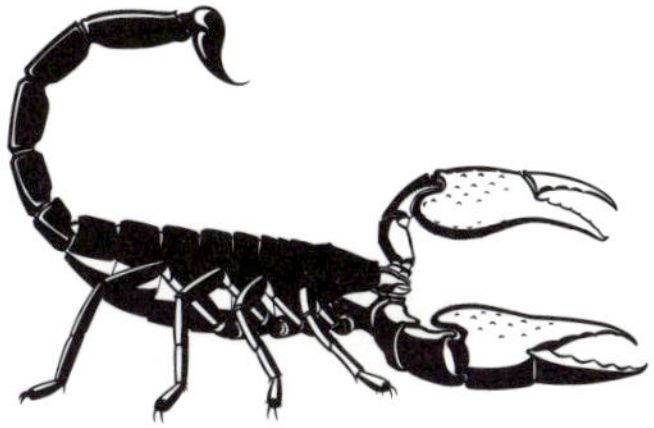

Scorpion

regions, they avoid the heat of the day, which they cannot endure, in this way. A scorpion exposed to heat dies quickly and the body curls over so much that the venomous sting ends up over the foreparts. This led to the erroneous belief that scorpions kill themselves with their own sting in the heat.

11 **Tailed Whip Scorpions** (Uropygi) **and Tailless Whip Scorpions** (Amblypygi)

These groups, which occur only in the tropics, have pedipalps modified into powerful claws for seizing prey, while the first pair of cursorial legs is modified into long organs of touch. Large tailless whip scorpions have tactile legs with a span of up to 30 centimetres (displayed specimen), although most species are smaller. They live under wood and leaves, and go out hunting at dusk or by night. They locate their prey with the tactile legs, then seize it with a lightening quick thrust of the pedipalps which snap together clenching in the prey.

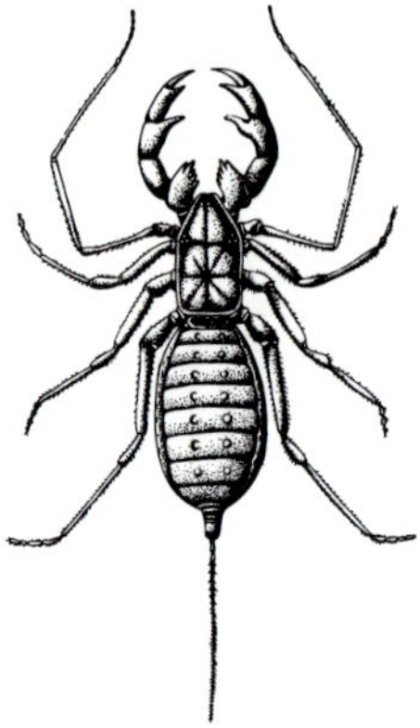

Tailed whip scorpion

Tailed whip scorpions are also nocturnal predators which conceal themselves in wood or leaves, and can even dig themselves funnels in the ground. Glands opening into a thin abdominal appendage secrete a very effective substance for defense against their enemies: it contains acetic and caprylic acid and causes burning pains. With their short, thick pedipalps, tailed whip scorpions can also deliver a powerful pinch.

11 **Sun Spiders** (Solifuga)

Sun spiders have strikingly large chelicerae which point directly forwards and can crush even quite large prey. The forepart is segmented so they can erect it and raise their chelicerae to take on an imposing threatening posture. The first pair of legs is no longer used for locomotion but for touching so they are more slender than the other legs. Sun spiders live in desert regions and are nimble nocturnal predators.

Sun spider (Solifuga)

Harvestmen (Opilones)

11

Amongst the species native to Germany, it is above all the long-legged species that are generally known. They climb around quickly in grass, plants and shrubs or often sit pressed flat on house walls and stone fences. Harvestmen can be easily distinguished from long-legged spiders since they have no "wasp waist", their body forming a single roundisch unit. A number of tropical species (Laniatores) have much enlarged basal sections of the fourth leg pair. Some species are very small (Cyphophthalmi, about 2 millimetres) and are shaped like mites. The Trogulidae are broad and flat, resembling bugs. The Ischyropsalididae feed exclusively on snails, which their tear apart with their large chelicerae.

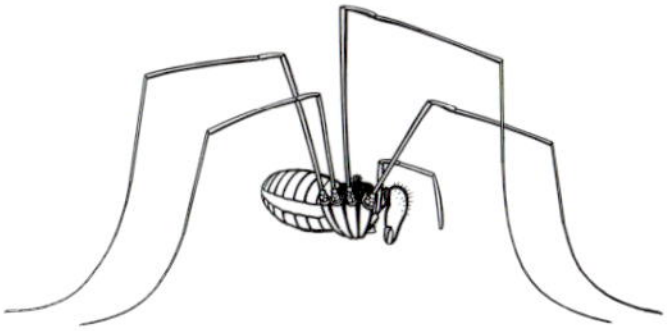

Harvestmen

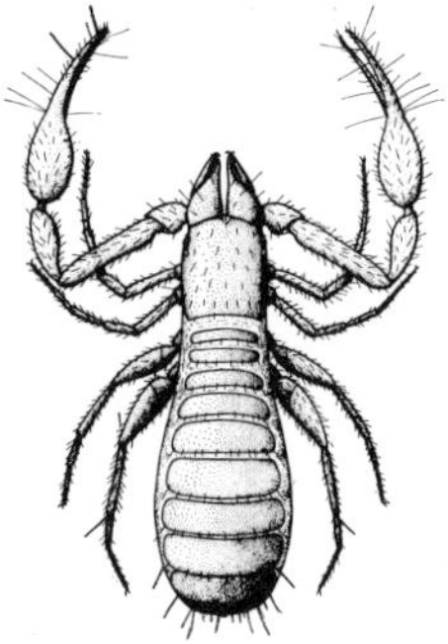

False scorpion

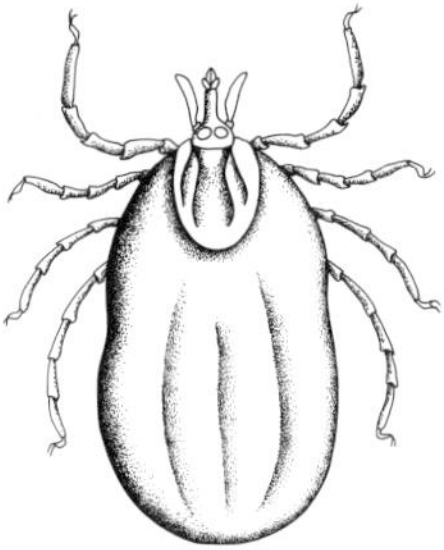

Sheep tick

False Scorpions (Cheloneti)
As the name indicates, the pseudoscorpions look like small scorpions without the abdominal appendage and venomous sting. The differences in other details are still more far reaching. At the most, pseudoscorpions grow to a length of 7 millimetres, most species being much smaller. They live mostly in cracks, where they can still move about thanks to their flat bodies, and under leaf litter, stones, wood, bark, in birds' nests, mammals' burrows and insects' nests (ants, bees). They fasten themselves onto the nest builders and have themselves carried out of the nests and so attain the dispersal of their species. Several species live in animal stalls and old buildings. The house scorpion ist often found in bookcases where they eat mainly minute booklice.

Mites (Acari)
There are about 10,000 species of mites which makes them the largest group of the Chelicerata next to the spiders. They are all very small, growing from between 0.08 millimetres (smallest of all arthropods!) to about 3 millimetres. Just because they are so small, a large number of possible modes of life are open to them and they are found almost everywhere. Some species have invaded freshwater and the sea. In addition to carnivores, carrion and waste eaters, there are plant eaters and suckers, which can become pests in times of mass incidence, damaging growing and stored crops. Many species are skin parasites (e.g., follicle mite), others are blood suckers (ticks) which can spread dangerous diseases. The female tick (the indigenous *Ixodes ricinus* is well-known) has a very elastic body wall and can expand up to centimetre size when it has sucked itself full.

Room 205, Dioramas (Habitat Groups)

205

The dioramas attempt to depict animals in their natural habitat. Some of them show mammals which used to be distributed over large areas of Europe during the Ice Ages, but are now found only in the far north or Inner Asia (reindeer, musk-ox, Saiga antelope), others show rare, almost exterminated Central European mammals (wisent, brown bear, wolf, ibex), and European mammals which, though still widely distributed, are practically never seen in their natural habitats by city-dwellers (red deer, roe deer, chamois, wild boar, fox, etc.) Some of these dioramas were created by the noted Frankfurt painter, Wilhelm Lefèbre in the years after the Second World War. The background of most habitat-groups represents typical landscapes from the environs of Frankfurt.

The reindeer and musk-ox lived in our latitudes during the last Ice Age. The reindeer, a species the females of which also have antlers, still live in the tundras and forest areas of northern Europe, Asia and North America. It is used as an easily tended domestic animal by certain nomadic peoples who follow it along its long seasonal migrations. The musk-ox is a wild cattle species well-adapted to the cold in its habitat in Arctic North America and Greenland.

The Saiga antelope is a steppe-dweller now found mainly in Inner Asia. During the Ice Age herds of them also used to wander about the loess steppes in Central Europe.

The wisent, a species of wild cattle, was once widely distributed over Europe and Asia. It was even more threatened with extermination than its North American cousin, the bison, and was saved only by being bred in zoos. Together with the aurochs, the forerunner of our domestic cattle, the wisent was an important hunting animal to our ancestors. The diorama shows

Saiga antelope

Wild boar

205

Elk

Beaver

wisents in the Schwanheim Forest near Frankfurt. The ibex and bearded vulture used to inhabit large areas of the Alps. While bearded vultures are now seen only rarely in the Alps, ibexes are kept in some alpine reserves as the remains of a once very diverse animal world. The brown bear belonged to that same world, and has managed to survive until today in the Italian Alps, though probably even this refuge is now threatened.

Our forefathers were so well aquainted with the wolf, badger, fox and bear that they became main figures in many animal fables. The wolf is now only found exceptionally in Central Europe: badgers and foxes, though not rare, are hardly ever seen because of their mainly nocturnal mode of life. Due to the danger of rabies, their numbers are kept small.

Wild boars mostly live socially in herds, preferably in moist areas: the diorama shows a family of wild boar in the bottomland forest of the Kühkopf near Frankfurt. Chamois and marmots are characteristic alpine animals. In some places they have become so accustomed to humans due to tourism that they can be observed easily. The alpine chough is to be seen near any mountain hut or rest place where it can find food in the waste. The largest of the European deer, the elk, is now only found in the north and northeast of Europe, where it inhabits sparse deciduous forests, swamp forests and marshlands. The diorama depicts elks and hooded crows on the Kurische Nehrung. There are large numbers of red deer in European highlands, while the roe deer, the most graceful and frequent indigenous deer species, is to be found almost everywhere in Germany.

Beavers, which were once widely distributed in Central Europe, as many place names testify, can now find hardly any suitable biotopes here where they could build their dams and lodges. The gray heron is suffering a similar fate and its breeding colonies have almost disappeared from Germany.

Room 206, Marine Invertebrates (Permanent Exhibition under Construction)

Basket-star *Gorgonocephalus*

A large number of very different types of animals live exclusively in the sea, have produced an enormous diversity of forms, and play an essential part in the overall balance of marine biotopes. All marine invertebrates which could not be taken into account in other rooms will be exhibited here.

Unicellular Organisms
Huge numbers of shell-bearing unicells, the radiolarians and foraminifers, live in the sea. They both belong to the Rhizopoda group, which

206

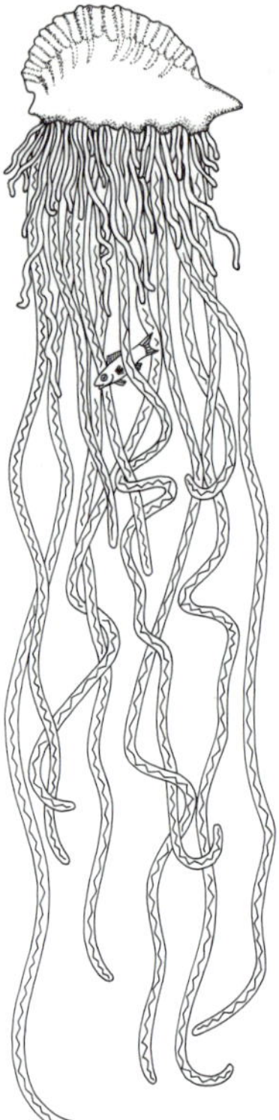

Portuguese man-of-war

also includes the amoebas. The aesthetically extremely attractive skeletons of the radiolarians consist of silicic acid and occasionally of strontium sulfate, those of the foraminifers of lime. These hard parts of dead unicells form deep layers of sediment on the seabed (see Room 7).

Sponges (Porifera)

The forms familiar as bath sponges are the horn-like fibers which make up the supporting elements of the sponge's body. Water swirls into the interior through innumerable small pores where fine particles of suspended matter are filtered out for food, then the water passes out again through the large openings. The bath sponges are found in depths still accessible to divers, while the bizarre hexactinellids live at great depths. Their skeleton is constructed of clear silica needles. There are about 5000 species which inhabit all regions of the sea and some freshwater biotopes as well. Chionids contribute to the breaking up of solid masses of limestone, especially in coral reefs. Fossil sponges are exhibited in Room 7.

Cnidarians (Cnidaria)

There are two very different forms of cnidarians: polyps, which are mostly sessile, and medusae, which swim around freely by rhythmically contracting their jelly-like bell, and catch small animals for food with their tentacles. The brown jelly-fish, Rhizostomeae and blue jellyfish are large species; the latter can grow up to a diamter of 2 metres.

The bizarre siphonophores are not medusae at all, but rather swimming polyp colonies composed of various types of polyps which fulfill various functions according to their different forms: feeding, swimming and prey-catching. The Portuguese man-of-war (*Physalia*) is known and feared for its enormously strong venom; its tentacles can be up to 50 metres long when extended to full length.

Horny coral (*Melithea ochracea*)

The hydrozoan polyps are mostly very small and form a sort of lawn on stones and wood, and even on ship cables. Many polyp species form medusae by budding. These medusae swim around freely and reproduce sexually, their larvae becoming polyps again.

The order of Anthozoa includes the sea anemones (actinians) and above all the many different corals. The stony corals (Scleractinia) forming their colonies in the cooler seas and deeper waters are small and only a few species, such as *Dendrophyllia*, which grows to several metres in height, become very large. The best known corals are the tropical reef corals. Under favorable conditions they grow very rapidly and their dead parts are turned into solid masses of limestone by calcareous algae so that reefs can quickly fixate

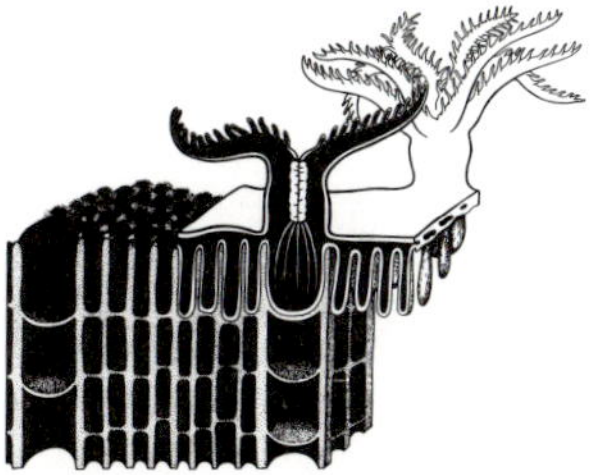

Diagram of the blue coal (*Heliopora*)

West Indian starfish *Oreaster*

enormous amounts of calcium carbonate and grow into huge embankments. Coral reefs offer a biotope for innumerable species of animals and plants including other anthozoans such as the alcyonarian corals and horny corals on display here. The latter group, many species of which also form colonies in deeper waters, includes the Mediterranean red coral often used to make jewelry or decorative art.

Bristle Worms (Polychaeta)
There are a large number of species of aquatic bristle worms, some of them swim freely, others dig into the ground, or are entirely sessile. The lugworm, which lives in very dense populations in the mudflats of the North Sea, is a familiar example. Many sessile polychaetans, such as the spiral

tubworm of the Mediterranean, have a broad crown of fine tentacles to catch suspension particles for food. Free-swimming polychaetans, like the tropical paloloworm can sometimes occur in huge numbers.

Echinoderms

The five-armed, radially symmetrical starfish, brittle stars, sea urchin, and the secondarily bilaterally symmetrical sea cucumber inhabit nearly all marine biotopes. Sea lilies are sessile forms whose fossils are important; one of the most beautiful fossil plates from Holzmaden is on display in the entrance hall.

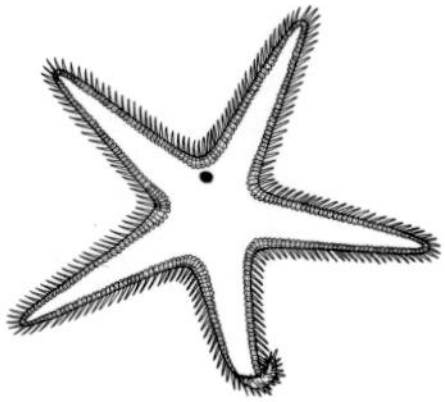

Starfish

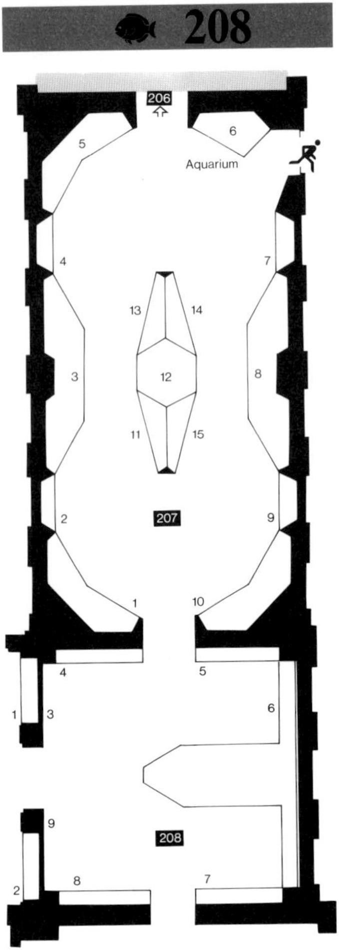

Room 207–208, Fish Gallery (Permanent Exhibition under Construction)

The circuit through the fish exhibition begins in the staircase outside Room 208, where there are two showcases with information on early research activities of the Senckenberg Society and the Senckenberg Museum.

Research on fish became one of the main focusses of scientific work in the Senckenberg Museum very early on due to its famous explorer, Eduard Rüppell, who collected specimens mainly from the Red Sea and Northeast Africa. Some of Rüppell's original preparations and drawings document his work. Today, scientists from the Senckenberg Research Institute carry out studies on marine and freshwater fish, their modes of life and their biotopes in many parts of the world.

Room 208, Jawless and Cartilaginous Fishes

The vertebrates commonly called fish have so many different developmental levels that they have to be divided into at least three classes: jawless fishes, cartilaginous fishes, and bony fishes. So far, no fossil evidence is available for the phylogenetic origin of the vertebrates. The first animals with backbones (Chordata) were probably small delicately built, fish-like creatures originating at the beginning of the Paleozoic era or even earlier but not preserved as fossils.

Lancelet (*Branchiostoma*)
The lancelet, which nowadays occurs in the North Sea and sandy Atlantic coastal waters, largely corresponds to scientific hypotheses about the most archaic chordates. Its internal skeleton consists only of a flexible skeletal rod called a notochord. They use a simple fin margin and

Palenose parrotfish (*Scarus psittacus*) collected and prepared for exhibition by Rüppell

tailfin to swim but entirely lack a head, brain, cartilage, and bones. The lancelet can be regarded as a living fossil still possessing the most important anatomical features of the Chordata which originated in the Paleozoic era. It is probably a very close relative of the evolutionary predecessors of the vertebrates.

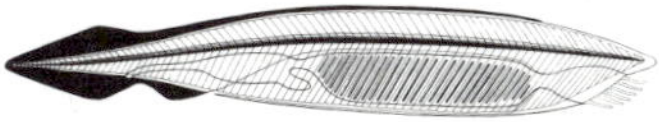

Lancelet

Phylogenetic Tree of the Jawless Fishes and Fish 4

This attempt to reconstruct the evolutionary process is based exclusively on fossil finds and on present day jawless fishes and fishes.

The ostracoderms (Ostracodermi), classified as jawless fishes, are the oldest group for which there is fossil evidence, and come from the Silurian and Devonian periods of the Paleozoic era. Nevertheless, they have already reached an advanced evolutionary level. Their precursors are completely unknown and can only be the subject of hypotheses. The same holds for the origins of the modern hagfishes (Cyclostomata).

The placoderms (Placodermi), appearing from the beginning of the Silurian period onwards, had a complicated bodily structure. Their origins are just as hypothetical as the question of whether they were forerunners of the cartilaginous fishes which appeared somewhat later. The same is true of the acanthodians (Acanthodii) which all lived at almost the same time during the Devonian period of the Paleozoic era. Whereas until recently, they were considered to be ancestors of the cartilaginous fishes, they are now thought to be precursors of the bony fishes. The prehistory of the lung fishes (Dipnoi), which belong to the bony fishes, is also unknown. The lobe-fin fishes (Crossopterygii) are thought to be the ancestors of the terrestrial vertebrates. The ray-finned fishes, which form the genuine bony fishes, can be divided historically and phylogenetically into a series of stages and developmental levels. The lowest level is occupied by the chondrostean fishes (Chondrostei) which originated in the Devonian period and reached their greatest diversity of species during the Carbonaceous period. Their direct descendants are the Old World sturgeons and the paddlefishes of North America and East Asia. The next level is occupied by the holostean fishes (Holostei); the largest number of forms they ever developed lived during the Mesozoic era and their modern descendants are the gars (*Lepisosteus*) and the bowfin (*Amia*).

The genuine bony fishes were the next, and last stage so far. Their evolution took place mainly in the Cretaceous and subsequent Tertiary periods, and at present they have the highest diversity of species. The many different forms range from the eel to the opah, and from the carp to the sea horse and trunk fish. They inhabit all aquatic biotopes in the sea and inland from mountain torrents to oceanic deeps of 10,000 meters completely without light; some species have even managed to adapt to a certain degree to life in atmospheric air. The genuine bony fishes are not only the vertebrates with the largest number of species at present, but they also have the greatest diversity of forms and can be very adaptable to various environments and biotopes.

Hagfishes and Lampreys (Cyclostomes)
Lamprey eels (Petromyzonida) and slime eels (Myxini) are present day Cyclostomata which belong to the jawless fishes. Their most important chracteristic is that they have no real jawbones which links them with the ostracoderms although they look completely different. All modern Cyclostomata lead a semiparasitic life.
The lampreys (the specimen on display is a river lamprey, *Petromyzon = Lampetra fluviatilis*) have a round sucking mouth with gripping teeth and a rasping tongue. They have a single nostril on the upper side of the head, then, on either side of the head, behind the eye seven pore-shaped gill openings. They attach themselves firmly to other fish with the sucking mouth and teeth, rasp a hole in the skin and gradually suck their victim dry. It is now generally accepted that the lamprey is related to the Cephalaspidomorphi species of the ostracoderms.
Slime eels, like the common hagfish (*Myxine glutinosa*), have a slit-shaped mouth and whiskers. Depending on the species there are between 1 and 16 gill openings along the side of the body; the North Sea common hagfish has only one pair. They are regarded as being related to the Pteraspidomorphi species of ostracoderms.

All the following fish groups belong to the jawed animals (superclass, Gnathostomata).

Cartilaginous Fishes (Chondrichthyes)
The cartilaginous fishes form a class by themselves to which the sharks, rays and chimaeras belong. Their internal skeletons are composed entirely of cartilage. Even the skull is cartilaginous and has genuine jaws. By contrast, the dermal skeleton is composed of dermal denticles. While sharks have dermal denticles thickly covering

4

River lamprey

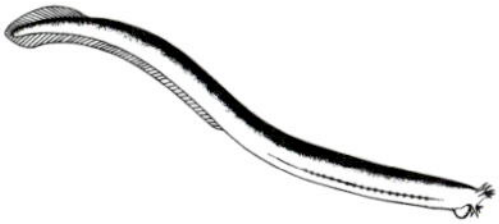
Slime eel

their whole bodies, rays usually have them only as longer spines on the middle of their backs, and on the upper side of their tails.

5

Sharks

The internal skeleton, the fins and the skulls of sharks consist of cartilage only (see specimen skeleton). The vertebral column is fully developed. The tailfins are asymmetrical with the tail vertebrae always ending in the upper fin. The 5–7 gill slits on each side of the rear of the head are connected to the anterior section of the gut, the pharyngeal gut. The mid-gut has spiral membranous folds which increase the internal surface of the gut and reinforce and accelerate the digestive processes with numerous glands. Sharks have no swim bladder though the very fatty liver may also function as a buoyancy organ. While many sharks live in open waters, numerous species prefer the seabed as their biotope. The forms of sharks' teeth are related to their feeding habits: species which mainly feed on other large fish and aquatic mammals have a single row of saw-like teeth, while those that eat hardshelled animals have several rows of teeth united to form grinding plates.

Dusky shark

Reef shark

Two developmental levels can be distinguished in present day sharks. The archaic sharks with 6–7 gill slits include the cow sharks (Hexanchiformes) with the frilled shark which has 6 gill slits, and the horn sharks (Heterodontiformes) with the Californian horn shark.

The true sharks (Lamniformes) and dogfishes (Squaliformes) belong to the more highly developed sharks with only 5 gill slits. Amongst others, the true sharks include the nurse sharks with pistonshaped bodies, the spotted dogfish, the combtoothed shark and the mackerel shark (see further the shark species on the front wall of the room), as well as the basking shark (above the showcase) and the hammerhead (hanging in the room).

Display of sharks. In front of it sawfish (*Pristis microdon*), angel fish (*Squatina squatina*) and common skate (*Raja batis*)

The dogfishes are represented by the spiny dogfish which is economically important in Germany and is sold as "sea eel" and "thorn fish".
Another shark group consisting of only a few species is the saw-sharks (Pristiophoridea) – not to be confused with the sawfish – which have 5–6 gill slits on their sides.
Angel sharks (Squatinoidea) are also counted as being sharks. The angel fish (prepared specimen on the iron grid to the left) has a flattened body like a ray, but the gill slits are on the sides of the head in front of the pectoral fins.

Mako

208

The basking shark reaches a length of just 14 metres. It lives in the Atlantic and Mediterranean, but also occasionally in the North Sea, and feeds mainly on free-swimming crustaceans and plankton. The specimen displayed here (above the showcase) is a young one.

6 There are casts of several striking shark species mounted on the front wall and hanging from the ceiling.

Sharks which inhabit the high seas often have externally nearly symmetrical tail fins, while in species which stay in shallow coastal waters or even on the bottom , the upper lobe of the tail fin is always larger than the lower one.

The white shark has a powerful body and can grow to a length of 10 metres. It is one of the most feared sharks. Although smaller specimens have even been caught in the North Sea and the Baltic, it is dangerous to humans only in tropical zones. The mako from the Atlantic rarely grows longer than 3 metres and is a typical deep seas predators which tears tuna and sword fishes apart.

The reef shark (length: up to 2 metres) is a characteristic inhabitant of tropical coral reefs.

The closely related dusky shark occurs in the Atlantic in the environs of the Canary Islands and Madeira. If feeds mainly on fish.

The sand tiger shark (length: 3 metres) is a sea-bed-dweller in tropical and subtropical seas. When it wants to swim in open water it first swallows some air to provide itself with the necessary buoyancy. The upper lobe of the tail of the thresher (hanging on the ceiling of the room) is longer than its body. It hunts in packs in tropical and moderate seas, rounding up schools of fish by threshing its tail and driving them into dense clumps so as to be able to prey on them more easily.

The flattened head of the hammerhead with eyes at the ends of its lateral extensions is an unusual form (hanging on the ceiling of the room).

White shark

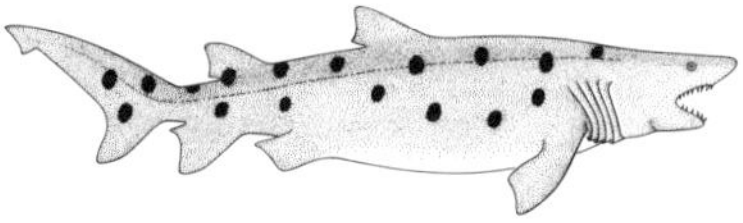

Sand tiger shark

Thresher

Hammerhead

Hammerheads live mainly in coastal waters and are found in all tropical and subtropical seas.
Despite their elongate bodies, sawfishes belong to the rays, and, like them, have 5 gill slits on the underside of the body, and the foreparts of the pectoral fins fused together with the sides of the head. It uses its long sawlike snout with rows of "teeth" on each side mainly to rummage around in the seabed looking for flatfishes and crustaceans on which it feeds. Its real teeth are in the form of pavement-like plates. The common sawfish reaches a length of up to 5 m (mounted on an iron grid on the floor in the middle of the room).

Rays

7

These cartilaginous fishes also have skeletons composed entirely of cartilage. Their bodies are normally compressed to flat disks and the tail fins only weakly developed. The five gill slits are always located on the underside of the body. The pectoral fins mostly merge into a unit with the body and form the locomotory organs. They, too, possess a spiral gut. With a few exceptions, rays live on the seabed.
Only the head and anterior of the guitar fish's body are flat, the rest of the body being cylindrical. This double form affects its way of swimming: while the pectoral fins of the anterior body flap up and down like a true ray's, the posterior part executes tail strokes sideways like a shark. In contrast to the true rays, whose flattened sides are used exclusively for locomotion, the electric rays have large electric organs in their bodies so that they can swim only with the aid of their tail fins. They can emit electric discharges measuring up to more than 200 volts and 2000 watts either in defense or to stun their prey.
The skates also have a flattened body. The back and upperside of the tail often have small spikes or humps which, however, are not poisonous. The wavelike movements of the sides of the body provide locomotion.

Ray

Stingrays live mostly in the subtropics and have long saw-toothed venomous spines on the section of the tail stalk near the body to defend themselves against enemies.

The sides of the eagle ray's body are broadened like wings so that they might be said to fly through the water. Nevertheless, they feed on fish, crustaceans and mussels from the seabed which is why they have teeth plates. Manta rays or devil fishes also have "wings", but they feed on small prey which they shovel into their mouths with two appropriately shaped appendages on their heads. Their teeth are only small.

Chimaera

7 **Chimaeras**

Because of their purely cartilaginous skeleton, the chimaeras are also classified as cartilaginous fishes, but form a separate subclass due to strongly deviating anatomical characteristics. While sharks and rays generally have 5 pairs of gill slits, chimaeras have only one gill opening on each side. They do not have a large number of teeth but two pairs of teeth plates in the upper jaw and one pair in the lower jaw. Their tails end in whipshaped points and are not used for swimming which they do with the aid of their paddle-shaped pectoral fins. One representative of the chimaeras is the ratfish, an inhabitant of the North Sea and the Mediterranean.

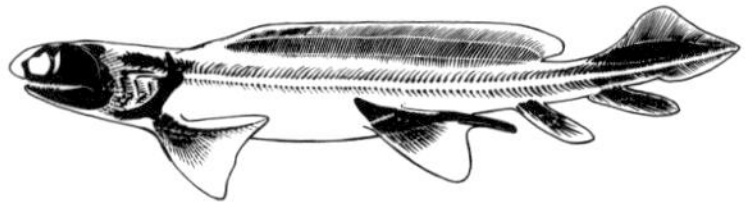

Reconstruction of a fossil freshwater shark

8 **Freshwater Cartilaginous Fishes**

Although cartilaginous fishes occur almost exclusively in the sea, one species of sharks, at least 14 species of stingrays, and some sawfish species live either sometimes or permanently in freshwater.

The bull shark, a comb-toothed shark, which reaches lengths of more than 3 metres, prefers shallow coastal and lagoon biotopes. It occurs in all 3 oceans and swims for distances of up to 1000 km into large rivers such as the Amazon in South America, the Zambezi in Africa, and the Euphrates, Tigris and Ganges in South Asia.

There were cartilaginous fishes in freshwater in early geologic eras as well. For instance, the fossil shark *Orthacanthus senckenbergianus* on display lived in freshwater lakes which had formed in the Saar-Nahe area in western Germany about 275 million years ago. The fossil plate on display, the most beautiful specimen found so far, dates from the New Red stage.
The freshwater stingrays have a considerably lower concentration of urea in the blood than all other cartilaginous fishes as an adaptation to their freshwater biotope. They are widely distributed in the rivers of South America, while the one sawfish species living permanently in freshwater is found only in the rivers of northern and northwestern Australia.

Room 207, Bony Fishes (Osteichthyes)

The class of bony fishes contains the true bony fishes and a number of ancient forms which can be called "living fossils". The degree of development and so the phylogenetic stage are recognizable in particular by the form of the skeleton and the structure of the scales: scales coated with shining ganoine (ganoid scales), an ossified skull, and asymmetrical, vertically oriented tail fins are phylogenetically ancient characteristics.

Polypteridae

The most ancient bony fishes still living today are the Polypteridae from tropical Africa. They are regarded as being descendants of the primeval Palaeonisci. Their ganoid scales are largely identical with those of the oldest bony fish, *Palaeoniscus*, and, like it, they also have an internal skeleton entirely of bone. Apart from the unusual formation of smaller additional finlets on the spiny ray of the dorsal fins, their pectoral fins have a supporting function and consequently a skeletal structure reminiscent of terrestrial ver-

1

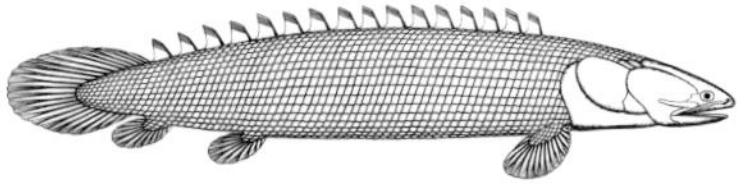

Polypterus

tebrates. Their swim bladders are modified into simple lungs so that they breathe mainly by surfacing and taking in atmospheric air.

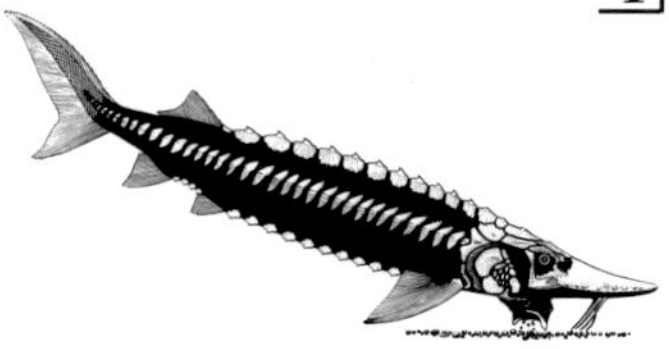

Sturgeon

1| **Cartilaginous Ganoids** (Chondrostei)

Sturgeons and paddlefish are Chondrostei and belong to the most ancient relic forms of the bony fishes. Their internal skeleton is still largely cartilaginous, and their vertebrae not yet fully formed so that their notochord is still well developed. They have, however, an external skeleton with an almost complete external skull, and bony plates set into the skin in longitudinal rows. Sturgeons live both in the sea and in freshwater. Usually they spawn in the central reaches of rivers but spend most or their life in the sea. Sturgeons used to be caught regularly in European rivers. Not only their flesh but also their roe (caviar) are considered to be delicacies. The paddlefish, with its characteristic spatula-shaped snout, is an inhabitant of North America. The sturgeons soak up their food from the bottom through extended jaws, whereas the paddlefishes filter plankton with gills while swimming with their mouthopen.

1| **Bony Ganoids** (Holostei)

The Holostei are also ancient, but somewhat more highly evolved bony fishes. They used to form a numerous group of various origin. Today, there are only two species still living in North and Central America (during the Createceous and Tertiary periods they also occurred in Africa and Europe). Their internal skeleton is largely ossified, but one of the groups still has primitive scales. This is the gar (*Lepisosteus*), which still has genuine ganoid scales and vertebrae with joints such as are found otherwise only in terrestrial vertebrates, but not in fish. In addition, its swim bladder is modified into a lung. Its closest relative is the bowfin (*Amia*) which has thin cycloid scales and an external skull with bony

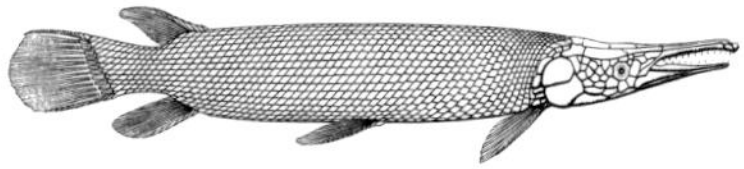

Gar

plates covered with a fine shining layer of ganoine. The Holostei were originally widely distributed because North America and Europe were located alongside each other in the Mesozoic era before the continents drifted apart.

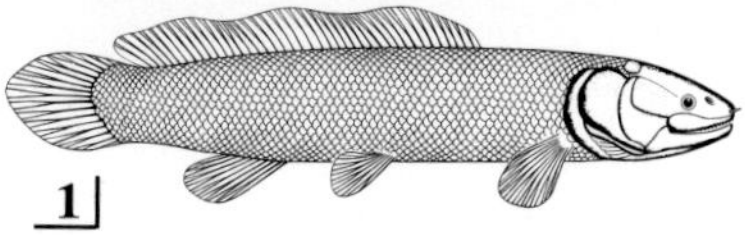

1

Bowfin

Lung Fishes (Dipnoi)

Lung fishes are a group which specialized very early on and now form a dead end in the phylogenetic tree. Their oldest relatives in the Paleozoic era were specialized in ways very similar to the modern lung fishes. One of these specializations in particular is that the teeth are united into grinding plates (see Room 6). Modern lung fishes are inhabitants of the southern hemisphere, with one genus each in the Amazon region, in African rivers, and in Australia. The Australian species, *Neoceratodus*, is most strongly reminiscent of the fossil forerunners. All modern lung fishes live in small waters and use their swim bladder as a lung either constantly or at times when the oxygen content of the water is low. The African species of the genus *Protopterus* can even encapsulate themselves in the mud when the water dries up completely and survive by estivating. The modern distribution pattern of the lung fishes is only explicable by assuming that there used to be one "primitive continent" comprising South America, Africa and Australia.

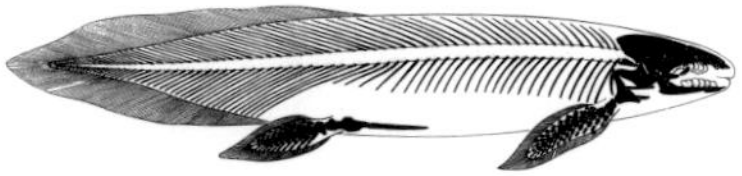

Skeleton of a fossil lung fish (reconstruction)

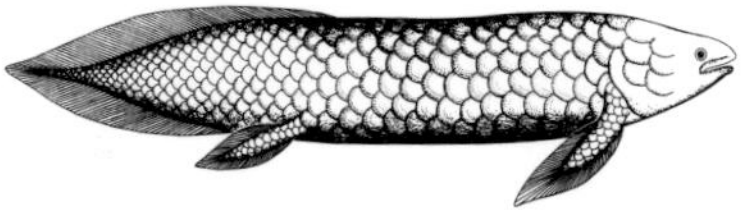

Australian lung fish

2

Lobe-fin Fishes (Crossopterygii)

The lobe-fin fish, *Latimeria*, on display as a life-size cast, is a living fossil; there is an original specimen in the scientific collection. This species was first discovered only in 1938, and so far only about 100 specimens have been found in a small area near Madagascar and the Comoro Islands. Its closest relatives lived during the Mesozoic era. Lobe-fin fishes from the Devonian period are of great evolutionary significance as predecessors of terrestrial vertebrates. The majority of lobe-fin fish species lived during the Paleozoic era about

Lobe-fin fish (*Latimeria chalumnae*), cast of a fullgrown specimen

300 million years ago, the last species dying out during the Cretaceous period about 70 million years ago, as far as is presently known. *Latimeria* is apparently the only species left over. It has almost the same anatomical features as its ancestors.

The skull consists of an anterior and a posterior half connected only by joints of muscles and tendons. This division of the skull into two blocks is regarded as an embryonic feature carried over into the adult stage (neoteny). The cranium, in particular, is only partially ossified. With the exception of the first dorsal fin, all fins have a stalk-like base which enables them to be revolved like propellers. The body axis consists of a tube of elastic tissue filled with fluid (500 milliliters in a 1.30 metres long fish), another feature regarded as neotenic. At the end of the axial tube, there is an additional small fin in the middle of the caudal

Skeleton of lobe-fin fish

fin, a feature found in many extinct lobe-fin fishes. The eggs contained in the body of the female are unusually large, weighing an average of 300 grams. They have no sort of protective shell whatsoever. These eggs develop inside the womb into the young which are born when they are fully formed.

True Bony Fishes (Teleostei) 3

These bony fishes make up the main proportion of modern fish. Their bony skeleton distinguishes them from the other groups. Although the true bony fishes are the modern representatives of fish, a lower, a middle and a higher evolutionary level can be distinguished. In the more ancient forms, the swim bladder is connected by a duct with the gut; moreover the pelvic fins are located on the belly, the fin-rays are soft-rayed and the scales are smooth-rimmed (cycloid). Members of the more primitive and generalized groups of teleosts are the bony-tongues and the tarpons, followed by the herrings, minnows and carps as well as the eels. In the highly evolved level, the swim bladder is closed, the pelvic fins are located on the breast or even on the throat, the anterior fin-rays are non-segmented spines, and the scales have minute spines on their posterior margins (ctenoid). This group includes the percoids and the flatfishes. There are, however, numerous transitional forms so that the groups cannot be distinctly separated from each other. In a number of cases both prepared specimens and their skeletons are displayed for purposes of comparison.
The arapaima from the Amazon belongs to the bony-tongues or osteoglossomorphs and is one of the more ancient types of teleosts. It is one of the largest freshwater fishes and reaches a length of about 4.5 metres. Another gigantic form is the tarpon which can grow to a length of more than 2 metres and a weight of more than 300 pounds, and is one of the most popular sport fish in America. Although it is a sea-dweller, it occa-

Arapaima

Conger eel

Carp

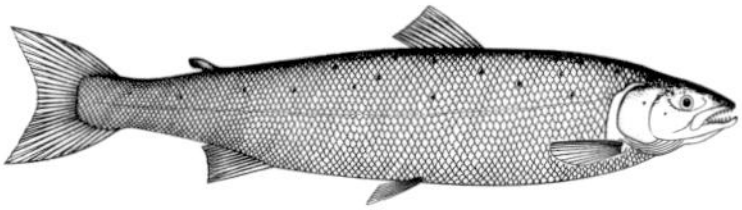

Salmon

sionally also swims into rivers. The conger eel from the North Atlantic also reaches a length of over 2 metres but always remains in seawater.

The Atlantic herring reaches a length of 40 cm and may live for 20 years. It lives on both sides of the North Atlantic and also in the North Sea and Baltic Sea. It is one of the most important food fishes. The shads, belonging also to the herrings and living in the eastern Atlantic and Mediterranean, migrate into rivers for spawning.

Even though cyprinid and catfish (Ostariophysi) species are very differently structured, they have one feature in common, the "Weberian apparatus", auditory ossicles connecting the inner ear with the swim bladder. Many representatives of this group are popular aquarium fish, for instance, the characins of which there are more than 500 species from Africa, Central and South America. The best known of these three groups is the carp which grows to a maximum of 1.20 metres and 20 kilograms making it both the largest of the cyprinids and the most important edible freshwater fish. Its original habitat is in the eastern Balkans and southern Russia. Only two of the total of 2400 catfish species from the tropical and temperate zones are found in Europe. These are the imported ictalurids and the European wels, which can grow to a length of 3 metres. The adults feed on fish, frogs, birds, and small mammals. The armored catfish with its powerful ossified skin is one of the tropical species.

The salmons and trouts, the codfishes and the anglerfishes are regarded as transitional forms to the higher teleosts. The salmonids include not only the different species of trouts and salmons, whitefish and graylings, but also the pikes. The most popular species of the codfishes is the Atlantic cod, which is one of the most profitable food fishes. The anglerfish or monkfish, which reaches a length of 4 feet, is one of the pediculates which owes its name to its special method of catching prey: it has a strongly armed mouth opening obliquely upwards so as to snap at the small fish attracted by its first ray of the dorsal fin which projects over its head like a bait.

Higher bony fishes

The fishes in the showcases 4 and 5 belong to the higher bony fishes, the spiny-rayed fishes or Acanthopterygii. They have, at least in most cases, both soft rays and spines in the fins. Furthermore the pelvic fins are placed far forward on the breast region.

4

Opah or moonfish

A striking higher teleost in this showcase is the particularly brightly colored opah or moonfish, a rare inhabitant of the high seas, sometimes also caught in the North Sea. Its skeleton is highly adapted to swimming with the pectoral fins. Like its relatives this species has a significant protractile mouth. Its closest relative is the greatly elongate and ribbon-like dealfish, an inhabitant of the deeper waters of the Atlantic and occasionally of the North Sea, which can reach a length of 10 metres.

The berycids have mucous cavities on the head and mostly well developed spiny scales. The northern red fish is a deep-water inhabitant of the northern Atlantic. The John Dory grows up to 70 centimetres or even 1 metre long and lives in the Atlantic and the Mediterranean.

5

Seahorse

The sticklebacks and tube-mouthed fishes are characterized by more or less elongated snouts, spines in the dorsal fin and bony plates in the skin. The sea stickleback has a long, thin body, 14–16 spines along the back and lives in the northeastern Atlantic including the North and Baltic Seas. Characterized by a very bizarre appearance are the tube-mouthed fishes. One feature which is common for all species of this order is the tubelike shape of the snout. Furthermore there is characteristic the complete or partial external armor of bony plates, which covers the body of these fishes. To these fascinating fishes belong the popular pipefishes and seahorses. An Australian relative of the seahorses is the weedy sea-dragon with the curious leaf-like outgrowths on head and body, which camouflage it in its haunts consisting of seaweed. Other relatives of the pipefishes are the razorfishes, cornetfishes and trumpetfishes, having a longitudinal body and also a tubelike snout. All these fishes eat small crustaceans and fishes. They suck their prey into the mouth by siphoning or pipetting the food.

The scorpionfishes or mailcheeked fishes have bony plates and spines on the head. Many species are characterized by venom glands on the base of the dorsal, anal and ventral fin spines. To this group belong the scorpionfishes or sculpins from the Mediterranean, the tropical lionfishes and turkeyfishes as well as the stonefishes which produce the strongest venom. This group also includes the rockfishes and gurnards (compare the showcase with Atlantic fish).

The order of teleosts with the largest number of species are the perchlike fishes or percoids, which include the perches, groupers, angelfishes, doctor fishes, the tunas and many others like the spiny eels. Swordfishes and sailfishes which also belong to the percoids, are gigantic inhabitants of the high seas. They owe their name to the sword-like beak formed by the upper jawbones. The sailfish can exceed 3.5 metres in length, and specimens weighing more than 500 kg have been caught.

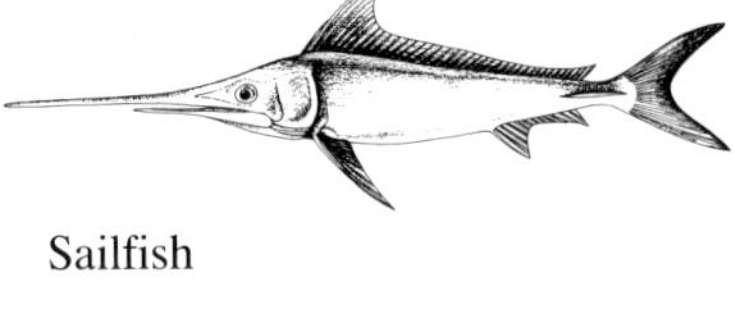

Sailfish

The flatfishes are offshoots of the perchlike fishes. They have a laterally compressed body which is oriented horizontally and has the eyes on the upper side while the lower side is blind and colorless. The turbot, halibut, brill, flounder, plaice, sole and many other species belong to this group, which includes the world's most valuable food fishes (compare also show-case 10).

The puffers and their relatives live mostly in the tropical and subtropical seas. But the ocean sunfish which has a striking shape and appears to be all head and body, lives in the tropical and temperate oceans including the northern Atlantic till Labrador and Norway. A specimen more than 3 metres long and weighing over 450 kg is on record. This species feeds mostly on jelly-fish, medusae, salpa, small squids and small fish. Other members of this order are the triggerfishes, pufferfishes, porcupine fishes, trunkfishes and related forms. They all have rather peculiar body shapes and small mouths with strong teeth. Most of them have bony plates, spines, or a leathery skin instead of scales. Many species are inhabitants of coral reefs.

Aquarium 6

An aquarium shows a section of a coral reef and contains vertebrates such as tubicolous polychaetes and corals as well as some small fish.

Fish in the River Main 8

Until the middle of last century, the Main in its natural state was one of the most abundant fish rivers in Europe. Then, between 1850 and 1860, as industrial pollution of its waters began, and from 1882 on, when canal and lock construction was initiated to turn the river into a commercial waterway, the animal population in the lower Main underwent a profound change.

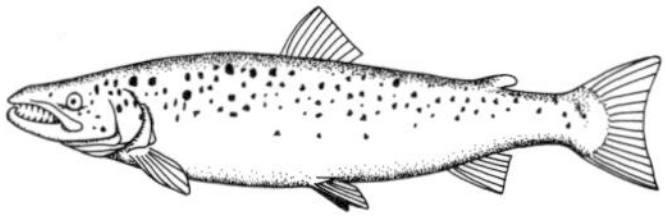

Salmon (Male with hook)

Originally, migratory fish such as the sea lampreys used to swim into the Main to spawn every spring.

The salmon, a bony fish, used to migrate in swarms upriver as far as the tributaries of the middle Main to spawn there. Also the shad, a species of the herrings, came up the Main as far as Bamberg just as regularly. Sturgeons came only very rarely into the Main and had to be handed over to the church when caught. The flounder, a well-known flatfish from the North Sea, used to migrate regularly into rivers, though not to spawn but rather in search of food. All

Case with fish of the River Main, former species (foreground), present day species (background left)

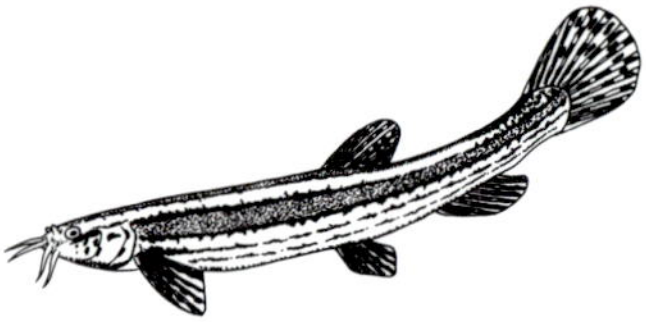

Loach

these fish species disappeared from the Main about 100 years ago. The only migratory fish still often found in the Main is the eel (these, however, are mostly eels released young into the middle and upper reaches of the Main, and not ones which have migrated upriver).
Whereas the local fish in the lower Main used to be mainly species which prefer fast-flowing, cool water with high oxygen content (particularly the barbel, nase, schneider, strömer, minnow, spined

loach, weatherfish, bullhead, and burbot), it is now chiefly stillwater fish which are found in this sluggishly flowing, very heavily polluted river: in particular fishes of the carp family such as the European roach, gudgeon, bleak, dace, chub, rudd, bream, common carp, and tench. The pike can find no spawning grounds in the lower Main. At the bottom of the showcase, there are models of cross-sections through the bed of the river as it used to be 150 years ago and as it is today. A comparison shows clearly that there used to be a large variety of different biotopes and spawning grounds for fish in the river and its tributaries, while the modern riverbed has become largely uniform and unecological due to canalization.

Perch

Migrations of Fish 9

There are many species which perform a variety of movements from one habitat to another, mostly for breeding or feeding. Astounding migrations in the ocean are carried out by varies tunas. The bluefin tunas have their spawning grounds in the Caribbean Sea where they appear in huge numbers in May/June, followed by a migration northward up the American Atlantic coast. During the summer months, they are present around Long Island and adjacent areas, and then move to the central Atlantic where they seem to winter in the deep water. Diadromous fishes which migrate between the sea and fresh water, are the eels and salmons. The spawning ground of the European and the American eel is an area of the Sargasso Sea in the western Atlantic where they seem to spawn in a depth of 400–700 metres. The eel larvae which are delicate, leaf-shaped and transparent, drift with the ocean currents for several months or even years to reach the North American or European sea coast. In the meantime they change from larvae to glass eels. As small elvers they ascend the streams and travel great distances upstream, where they grow up to adult specimens. After a freshwater life of 4 to 7 years the eels descend the streams and enter the ocean to migrate back to their spawning area in the Sargasso Sea. The Atlantic salmon spends most of its life in the feeding grounds of the northern Atlantic. For spawning the adult fishes migrate to the coast and swim up the rivers to their spawning grounds. While the adult specimens return to the sea after spawning, the juveniles have a freshwater life for about 1 year. Afterwards in the sea they grow up to adult specimens for 2 to 3 years and return to the stream for spawning.

10 **Fish of the North Atlantic**

This showcase contains the three main habitats of the ocean: the open sea, the sand bottom and the rocky coast. Each one has its special fish communities, the species display structural modifications of the body in relation to the conditions of life of the environment. The fishes of the open sea are streamlined and swift swimmers, such as tunas, mackerels and herrings. The standard color pattern of these fishes is the dark back, silvery flanks and a white belly. Fishes of the sand bottom are mostly light colored. They live on the ground such as flatfishes or habitually bury themselves in sand as the sandlance and weeverfishes. The colorful sea robins are partly protected by a cuirass of bony shields. They walk over the bottom by means of the 3 lower rays of the pectoral fins. The pogge is also covered with bony plates. It feeds mainly on small crustaceans. Fishes of the rocky coastal area mostly have a rather stout body, wide pectoral fins and are cryptically colored. Sea scorpions feed mainly on crustaceans, but also take small fish and mollusks. The sea catfish has very powerful teeth and feeds on crabs, hermit crabs, bivalve mollusks, sea-urchins, starfish and similar hard-shelled animals. The lumpsucker has a large sucker disc under the head and feeds on crustaceans, worms and jellyfish.

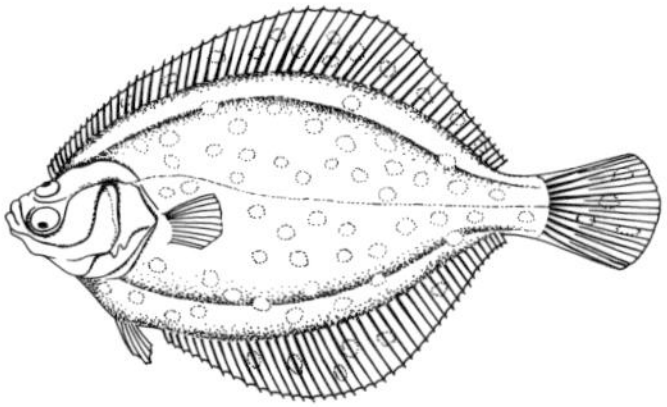

Plaice

The central showcase will contain a coral reef with a number of characteristic fish species. The surrounding showcases deal with special topics: poisonous and dangerous fish – special fish biotopes – distribution of earlier and of present-day fish – anatomy of the bony fishes. On the rear of this central block there is a screen showing three computer programs: The global water balance and water cycle, anatomy and functional morphology of the fish and the hydrological situation of the river Main.

12

The History of the Senckenberg Nature Research Society

Johann Wolfgang von Goethe suggested to the Frankfurt citizens to found a natural science society and to establish a "nature museum".	1815
On November the 22nd, the Senckenberg Nature Research Society, named in honour of Dr. Johann Christian Senckenberg (1707–1772), was founded.	1817
The first museum was opened at Eschenheimer Tor.	1821
Through the exploratory expeditions and scientific activities of Eduard Rüppell, Senckenberg joined the rank of world museums.	1820–1866
Dr. Otto Volger, paleontologist at Senckenberg, established the "Freie Deutsche Hochstift" and bought the Goethe House.	1857
The research collections were assigned to professional scientists.	from 1900
The present museum building, at that time situated on the periphery of Frankfurt, was erected and inaugurated. The exhibition museum was separated from the research collections.	1904–1907
Geographer Alfred Wegener introduced his famous theory of drifting continents to scientists and to the public for the first time at the Senckenberg Museum.	1912
The Senckenberg Nature Research Society donated the zoological, the geological-paleontological and the mineralogical institutes to the newly established University of Frankfurt.	1914
A research institute for marine geology and marine biology, "Senckenberg at the Sea", was founded in Wilhelmshaven.	1928
The collections of the Senckenberg Museum were stored in approximately 50 warehouses in the country side for protection during the war.	1939–1944
As 60% of the building were destroyed during the years 1944/45, the restauration was initiated and the stored collections were returned.	1946
Nominally, the Senckenberg Museum was divided into the Natural History Museum and the Research Institute.	1952
Senckenberg was recognized as a widely influential research institute of international importance by inclusion into the budgets of the Federal States, the so-called "Königsteiner Treaty", later changed to the so-called "Blue List", a treaty between the Federal Republik and the Federal States.	1954
The renovation of the exhibition museum commenced.	1962
The "Senckenberg School" for technical assistants was officially established.	1966
Celebration of the 150th anniversary of the Senckenberg Nature Research Society.	1967
Establishment of a new research department of paleanthropology.	1968

Inauguration of out-station "Lochmühle", section for cryptogams, at Biebergmünd (Spessart).	1969
Inauguration of the temporary environmental exhibition "Umwelt 2000".	1970
Extension of the research institute for marine geology and marine biology "Senckenberg at Sea" in Wilhelmshaven.	1965–1972
Out-station at North Norway; a laboratory boat for hydrological research at River Main and Rhein was put into service.	1975
The research ship "Senckenberg" – suitable for use in the North Sea – was put into service for the research institute for marine geology and marine biology at Wilhelmshaven.	1976
A former factory building in Kuhwaldstraße in Frankfurt am Main was bought and transformed into a geological-paleontological research institute of Senckenberg.	1978